Multiple Choice Questions in Basic and Clinical Physiology

Dom Colbert

Department of Applied Physiology
University College, Galway, Ireland

Oxford University Press, Walton Street, Oxford OX2 6DP

Oxford New York
Athens Auckland Bangkok Bombay
Calcutta Cape Town Dar es Salaam Delhi
Florence Hong Kong Istanbul Karachi
Kuala Lumpur Madras Madrid Melbourne
Mexico City Nairobi Paris Singapore
Taipei Tokyo Toronto
and associated companies in
Berlin Ibadan

Oxford is a trade mark of Oxford University Press

Published in the United States
by Oxford University Press Inc., New York

© D. Colbert, 1996

A catalogue record for this book is available from the British Library

Library of Congress Cataloging-in-Publication Data

ISBN 0 19 262 7368

Typeset by
EXPO Holdings, Malaysia

Printed in Great Britain by Biddles Ltd, Guildford

Preface

This book is intended to challenge and inform, to test and to evaluate and, finally, to prompt the student to search for further information in a standard, larger text. In most cases the answers are expanded to give extra information, but limitation of space means that answers are necessarily as brief as possible. Unless specifically stated, all questions refer to a normal human adult and each question (a–e) should be taken in conjunction with the specific headline statement that introduces that question. Note, too, that where a question contains more than one element a False answer will be the correct answer even if only one of the elements in the question is incorrect. Although we all dislike asking specific levels for body constituents some questions do just that. This is because a knowledge of certain range levels is part and parcel of applied physiology as it is of other disciplines. A few questions are also deliberately inserted to 'catch out' the inadvertent reader or those who do not read the material carefully.

Answers reflect the most obvious, generally accepted view at the time of writing, but are naturally coloured by my own views and experience. Sometimes the 'answer' will not closely match the question. This occurs when a brief True or False suffices to answer the specific question, so allowing me to expand and inform on closely related topics. Where the reader finds a variance between the given 'answer' and a standard text the likelihood is that I have used information gleaned from current review articles.

Apart from general and applied physiology suitable for all students, I have included specific questions aimed at testing those interested in anaesthesia, general medicine, surgery, obstetrics and gynaecology, psychiatry, and dentistry. These questions will be obvious to all readers.

Readers are invited to send in suggestions, criticisms, and ideas that will improve any future editions.

Galway
May 1996

D. C.

To those who have believed in me and supported me down
the years

Acknowledgements

I want to acknowledge the important contribution made by Dr Kevin Gormley, St George's Hospital Medical School, London, to the final presentation of this book. His depth of knowledge over a great range of subjects allowed him to make many constructive criticisms which greatly enhanced the final product. I also want to thank the staff at Oxford University Press for their advice, help, and encouragement throughout.

Contents

Managing MCQ tests

1. Know how many you need get right in order to get into the pass range (where possible). Assume you must score 60% taking negative marking into account.
2. At first only pencil in those answers of which you are sure. This may be all you need do to pass the exam.
3. Read questions carefully but do not read deeper meanings into them than is obvious from your first careful evaluation.
4. When in doubt, the correct answer may become obvious from reading through the question as a whole (all five parts).
5. Taking chances or guesses is almost certainly going to lose you marks.

A book like this can improve your success rate in MCQ examinations considerably. However, you must be completely honest with yourself when using it and should take the opportunity to expand your knowledge, in particular by group discussion of contentious issues.

Body fluids

Body water and osmolality (Qs 1–6)

1 *Body water:*

   a comprises about 55% by weight of the average 70 kg male.
   b is proportionately smaller in females than in males.
   c is proportionately larger in infants than in adults.
   d is chiefly in the intracellular fluid (ICF).
   e can be measured by calculating the dilution of sucrose in body fluids.

2 *The extracellular fluid (ECF) compartment:*

   a has a higher concentration of Na^+ than K^+.
   b is approximately equal to that of the ICF in the newborn.
   c has a higher concentration of ionic calcium than the ICF.
   d can be measured by estimating the dilution of radioiodine.
   e has a higher concentration of glucose than the ICF.

3 *The osmolality of plasma:*

   a is chiefly due to plasma proteins.
   b is almost all due to plasma sodium.
   c is the same as the osmolarity of plasma.
   d is the same as that of the ICF.
   e increases significantly in dehydration from vomiting.

4 *The 'anion gap':*

   a is the difference between plasma ionic sodium and ionic chloride.
   b should not be more than 1 mmol/l.
   c is also found in urine.
   d decreases markedly in diabetic acidosis.
   e increases in alcohol intoxication.

5 *About hyponatraemia:*

   a Chronic hyponatraemia is usually only clinically significant when the plasma Na+ falls to levels of about 135 mmol/l.
   b Symptomatic chronic cases should be treated by giving hypertonic saline.
   c It is a potential complication after a transurethral resection of prostate.
   d Hyponatraemia is a problem for formula-fed babies.
   e Hyponatraemia is typically caused by overtransfusion with normal saline.

6 *In hyponatraemia:*

   a Women are more likely to suffer brain damage than men.
   b Poor brain adaptation to the condition in children predisposes them to brain damage.
   c In acute cases, serum Na^+ should only be raised by about 5 mmol/l/h.
   d Acquired immunodeficiency syndrome (AIDS) should be suspected in the absence of other causes of water intoxication.

e the cause may be found in the practice of overtransfusing with 'normal saline'.

Body water and osmolality (Answers)

1 a **True.** Body water in an adult male ranges from 55% to 65% of body weight.

b **True.** Body water is relatively smaller in females (48%–58%) than in males.

c **True.** Body water is about 80% of body weight in infants.

d **True.** In the adult, ICF is approximately 28 litres and the ECF is approximately 15 litres.

e **False.** Dilution of sucrose will estimate ECF only.

2 a **True.** ECF sodium is about 135–45 mmol/l, ECF potassium is about 4–5 mmol/l.

b **True.** ECF and ICF volumes each approximate 40% of body weight.

c **True.** ECF level of ionic calcium is 2.5 mmol/l while that of the ICF is 0.5 mmol/l.

d **False.** Radioiodine tags plasma albumin. Therefore its dilution measures plasma volume only.

e **True.** Glucose is metabolized once it enters the cell.

3 a **False.** Plasma proteins exert approx. 1 mosmol/kg (25–28 mmHg) out of a possible 280–290 mosmol/kg.

b **False.** Sodium exerts osmotic pressure of approx 142 mosmol/kg, approximately 50% of the total.

c **False.** Osmolarity is expressed per litre while osmolality is expressed per kg. Use of the latter is preferred as it has the advantage of being temperature-independent.

d **False.** ICF is generally hypertonic to plasma because of the products of cellular metabolism. The cell membrane prevents rupture of the cell by translocating ions and organic osmolytes.

e **False.** Isotonic fluid loss occurs uniformly across ICF and ECF.

4 a **False.** The anion gap is plasma ($Na^+ + K^+$) minus plasma ($Cl^- + HCO_3^-$). It gives a measure of all the anions (proteins, lactate, etc.) that are not measured in the assay of plasma routinely.

b **False.** The anion gap ranges from 8 to 16 mmol/l.

c **True.** The anion gap is normally small. Can be used to measure urine $NH4_+$ (ammonium) ion output.

d **False.** The anion gap increases greatly due to organic anions, e.g. aceto-acetate.

e **True.** The anion gap increases due to the associated ketoacidosis.

5 a **False.** Asymptomatic chronic hyponatraemia may have a plasma sodium well below 135 mmol/l.

   b **False.** Hypertonic saline should be reserved for acute cases because of the potential dangers of heart failure due to an increase in Na^+ ions, or brain damage from fluid shifts.

   c **True.** Large absorption of hypotonic fluid occurs from the prostatic bed as, after surgery, the bladder is irrigated to inhibit clot formation.

   d **False.** Hypernatraemia is usually the problem due to the composition of formula feeds.

   e **False.** Hyponatraemia is typically caused by overtransfusion with dextrose/H_2O solutions, which are hypotonic *in vivo*.

6 a **True.** Brain damage in hyponatraemia is virtually confined to menstruating females and children, despite a similar incidence of hyponatraemia in adult males. The reason for this is unknown.

   b **True.** Other factors include hypoxaemia, cerebral vasoconstriction (induced by vasopressin), liver disease, and CNS abnormalities.

   c **False.** Na^+ serum levels should only be raised by 1–1.5 mmol/l to avoid circulatory overload and fluid shifts in the brain.

   d **True.** This is possibly due to associated mineralocorticoid deficiency in AIDS.

   e **False.** Isotonic saline contains 150 mmol/l Na^+ and 154 mmol/l Cl^-. It will not cause hyponatraemia. Note that D5-W (5% dextrose in water) is isotonic with plasma *in vitro* but hypotonic *in vivo* .

Potassium, magnesium, and calcium (Qs 7–18)

7 *About hyperkalaemia:*
   a A level of 8 mmol/l is always abnormal.
   b Hyperkalaemia is typically associated with acidosis.
   c Circulating catecholamines tend to lower elevated plasma potassium levels.
   d Hyperkalaemia may be treated by intravenous calcium gluconate.
   e A glucose–insulin infusion takes less than 5 min to lower plasma K^+ levels.

8 *Hypokalaemia:*
   a is arbitrarily defined as a serum potassium of less than 3.5 mmol/l.
   b is far less common in clinical practice than hyperkalaemia.
   c is a complication of all diuretic therapy.
   d increases the toxicity of digitalis.
   e in the range of 2–3 mmol/l is an indication for a monitored IV bolus of potassium even when the patient is asymptomatic.

9 In the syndrome of inappropriate antidiuretic hormone (SIADH):

a   urine osmolarity is greater than 300 mosmol/l.
b   hyperuricaemia is found.
c   oral urea is an excellent osmotic diuretic therapy.
d   oat cell carcinoma of the lung is a potential cause.
e   neurons secrete potassium and other osmotic particles in chronic cases.

10 Regarding magnesium:

a   Hypomagnesaemia is less common than hypermagnesaemia.
b   Hypermagnesaemia can cause 'magnesium tetany'.
c   Laxative overuse is commonly associated with hypomagnesaemia.
d   Aldosterone increases renal excretion of magnesium.
e   The plasma level is approximately 1 mmol/l.

11 Magnesium:

a   is an important vasodilator, especially in the cerebral circulation.
b   as magnesium sulphate is the drug of choice in established eclampsia.
c   as magnesium sulphate is the drug of choice in the prophylaxis of eclamptic seizures.
d   accelerates ischaemic damage in neurons.
e   is normally absent from the cerebrospinal fluid.

12 Calcium:

a   is one of the most important trace elements in the diet.
b   is the most abundant cation the body.
c   levels in the plasma rate-limit its absorption by the gastrointestinal tract (GIT).
d   is mostly reabsorbed from intraluminal fluid by the distal renal tubule.
e   excretion in the urine is increased by a high sodium diet.

13 Plasma calcium:

a   is physiologically active only in the ionic form.
b   decreases in hypoproteinaemia.
c   is rapidly decreased by therapeutic doses of ACTH.
d   rises as plasma phosphate rises.
e   is decreased by parathyroid hormone (PTH).

14 Movement of calcium in the body:

a   It is currently thought that Ca^{2+} channels in smooth muscle are all ligand-gated (L-gated).
b   Calcium antiport involves contrary movement of hydrogen ions.
c   The calcium–hydrogen pump is a Ca^{2+}, H^+-ATPase.
d   Intracellular organelles have receptors for the release of calcium.
e   Calcium combines more readily with plasma proteins if the pH falls.

15 Concerning hypercalcaemia:

a Acute cases are almost all caused by activation of osteoclasts which mobilize skeletal calcium.

b Malignant disease is the commonest cause of hypercalcaemia.

c PTH-related protein (PTHRP), which raises serum calcium, is secreted by normal healthy people.

d Polyuria is rare in hypercalcaemia.

e Plicamycin (mithramycin), an inhibitor of RNA synthesis, is useful in treating hypercalcaemia.

16 The treatment of acute hypercalcaemia includes:

a rehydration with normal saline.

b intravenous loop diuretics.

c thiazide diuretics.

d calcitonin in severe cases.

e glucocorticoids if hyperparathyroidism is present.

17 Malignant hyperpyrexia (syn. malignant hyperthermia):

a is due to a rare Mendelian autosomal recessive gene.

b is associated with an abnormality of the ryanodine receptor on the sarcoplasmic reticulum (SR).

c can be prevented if halogenated volatile anaesthetics and succinylcholine are avoided.

d is associated with an enhanced insulin response to glucose.

e is characterized by hyperpyrexia, muscle rigidity, and alkalosis.

Potassium, magnesium, and calcium (Answers)

7 a **False.** Transitory higher values can occur in exercise.

b **True.** Intracellular hydrogen ions displaces potassium from cells.

c **True.** Sympathetic nerves have the same effect via B_2 adrenoreceptors.

d **True.** $NaHCO_3$, dextrose/insulin, cation exchange resins, and potassium-losing diuretics are also used.

e **False.** Glucose–insulin takes 15 min to act and lasts for 1–2 h.

8 a **True.** Note that this is a biochemical definition and that much lower values are found in chronic hypokalaemia.

b **False.** Hypokalaemia is far more common in clinical practice.

c **False.** Diuretics such as spironolactone act to retain potassium.

d **True.** Both diminish the excitability of the myocardial cell membrane.

e **False.** The ICF/ECF ratio is normal in such cases. Rapid IV correction is harmful and will not correct total body potassium deficit.

9   a   **True.** Classically SIADH shows concentrated urine despite dilute plasma. Typically SIADH shows hyponatraemia, hypotonic plasma, and hypertonic urine.

     b   **False.** Hypouricaemia typical since H_2O retention dilutes the ECF.

     c   **True.** 30–60 g of urea/day is helpful in acute and chronic SIADH.

     d   **True.** Oat cell carcinoma typically produces ectopic hormones, especially ADH.

     e   **True.** Neurons lose KCl in SIADH by the opening of membrane K^+ and Cl^+ channels and by KCl cotransport. The secretion of organic osmolytes (polyols, amino acids, and methylamines) is a slower process.

10   a   **False.** Hypermagnesaemia is very rare.

     b   **False.** Hypomagnesaemia produces tetany of muscles.

     c   **True.** Laxatives are the most common cause of hypomagnesaemia.

     d   **True.** Most diuretics, alcohol, and ECF expansion also do.

     e   **True.** Normally 25% of plasma magnesium is ionic and parathyroid hormone (PTH) controls this. ICF Mg^{2+} is about 10 mmol/l.

11   a   **True.** Magnesium vasodilates, probably by increasing the synthesis of endothelial prostacyclin.

     b   True. $MgSO_4$ is superior to phenytoin or diazepam in preventing recurrent fits and in reducing death rates.

     c   **True.** $MgSO_4$ prevents the onset of fits both by being a cerebral vasodilator and an antagonist of the effects of the glutamate NMDA receptor (p. 232).

     d   **False.** Magnesium protects ischaemic cells, probably by preventing the entry of calcium ions.

     e   **False.** CSF Mg is about 1.1 mmol/l. The CSF/plasma ratio of Mg is about 1.5.

12   a   **False.** Calcium is not a trace element. An average good diet contains about 25 mmol (1 g) of calcium/day.

     b   **True.** Calcium contributes about 1 kg to the weight of a 70 kg man.

     c   **False.** Calcium absorption in GIT is rate-limited by mucosal calcium-binding protein (CaBP).

     d   **False.** 75% of calcium reabsorption occurs in the proximal convoluted tubule.

     e   **True.** Sodium enhances calcium excretion, possibly by causing a preferential reabsorption of sodium.

13   a   **True.** Note that the normal plasma calcium level is 2.5 mmol/l; 45% of which is truly ionized. 40% of Ca^2+ is protein bound and 15% is complexed with SO_4, PO_4.

     b   **True.** This is because less albumin is available to bind calcium ions.

     c   **False.** Although ACTH raises cortisol levels, cortisol is a fairly slow hypo-calcaemic agent since it works, in this case, by opposing vitamin D. Note, too, that cortisol demineralizes bone slowly.

     d   **False.** The product of plasma Ca^{2+} and PO_4^- is fairly constant.

e **False.** PTH raises plasma calcium. It demineralizes bone, retains calcium in kidney, and inhibits reabsorption of PO_4^- in kidney.

14 a **False.** Calcium also enters via voltage-gated channels (V-gated).

b **False.** Calcium is exchanged for sodium. Therefore this is indirectly powered by the sodium pump.

c **True.** Na^+, K^+-ATPase and H^+, K^+-ATPase are other cation pumps.

d **True.** Examples include ryanodine and inositol trisphosphate receptors.

e **False.** Calcium attaches with proteins better in alkalosis.

15 a **True.** Less common causes of hypercalcaemia include hypervitaminosis D causing excessive GIT absorption, and renal failure with an inability to excrete calcium adequately.

b **True.** Primary hyperparathyroidism is the second most common cause.

c **True.** Note PTHRP can also be secreted by malignant tumours.

d **False.** Hypercalcaemia blocks renal reabsorption of Na^+ and H_2O, thus leading to natriuresis.

e **True.** Given IV, plicamycin inhibits mRNA synthesis in osteoclasts.

16 a **True.** The resultant re-expansion of the ECF lowers the serum calcium at least by the degree to which dehydration raised it.

b **True.** Diuretics should only be given after rehydration. They inhibit calcium reabsorption by the kidney.

c **False.** Thiazides enhance distal tubular reabsorption of calcium.

d **True.** Calcitonin is a powerful osteoclast inhibitor and is used if plasma calcium is >4.0 mmol/l.

e **False.** Steroids are anti-vitamin D and anti-growth of neoplastic lymphoid tissue. Usefulness is confined to these causes.

17 a **False.** Malignant hyperthermia is caused by a rare Mendelian autosomal dominant gene affecting the long arm of chromosome 19.

b **True.** Ryanodine receptor controls uptake of calcium by SR.

c **True.** Thus susceptible persons can be safely anaesthetized. The role of stress and exercise is controversial.

d **True.** The supernormal insulin response is enhanced by free cytoplasmic calcium.

e **False.** Typically there is hyperpyrexia, rigidity, and **acidosis**.

The sodium pump and transmembrane potential (Qs 19–23)

18 *Concerning free intracellular calcium and cell death:*

a Free cell calcium rises in physiological cell death (apoptosis).

b Free calcium ions facilitate autodigestion.

c Calcium-blocking drugs can be used to slow cell death.

d Ischaemia can cause cell overloading with free calcium.

e Reperfusion of an ischaemic area may cause an increase in cell calcium.

19 The sodium pump (Na⁺, K⁺-ATPase):

a is confined to neural, muscle, GIT, and renal tissue.

b is confined to the outer surface of the cell membrane.

c is associated with the expulsion of two sodium ions and the entry of three potassium ions into the cell.

d contributes significantly to the resting membrane potential.

e carries an electric current across the membrane (electrogenic).

20 The sodium ATPase pump:

a The membrane density of pumps varies widely in different tissues.

b The pump is inhibited by thyroid hormone.

c The pump can work in either direction.

d The pump only functions when both Na^+ and K^+ are present across the membrane.

e Magnesium ions are necessary for its activity.

21 Concerning sodium pump activity:

a Cardiac glycosides and oubain only inhibit it if they are present in the ECF.

b Cyanide and dinitrophenol block the intracellular moiety of the pump.

c At rest, the sodium pump contributes about 45% of our heat needs.

d The sodium pump is responsible for the relatively high ICF potassium levels and relatively high ECF sodium levels.

e The sodium pump is one of several primary active ATPase cation pumps in the body.

22 The height of the resting transmembrane potential:

a is little influenced by the transmembrane gradient of Na^+ ions.

b is chiefly determined by the transmembrane gradient of K^+ ions.

c can be significantly altered by the movement of 1000 cations across the membrane.

d is little affected by the concentration of intracellular protein.

e is re-established by the electrogenic effect of the sodium pump just after depolarization.

23 Ionophores or ion channels:

a Ion channels are special lipid areas of the membrane that allow passage of specific ions.

b Their patency is regulated by: ligands, extracellular voltage changes, and Ca^{2+}.

c The Na^+ channel has three states: open, inactive, and closed (resting).

d A change of the Na^+ or Ca^{2+} channel from the inactive to the resting state is K^+ dependent.

e The K^+ channel has a less specific structure than other cation channels.

The sodium pump and transmembrane potential (Answers)

18 a **False.** In apoptosis cell membrane transport processes remain fairly normal so that intracellular calcium changes little.

   b **True.** Calcium ions activate autolytic lipases and proteases.

   c **True.** Although this is clinically not fully proven; for example, this occurs in myocardial and cerebral ischaemia.

   d **True.** Ischaemia also damages cells by causing anaerobiosis, acidosis, a decrease in ATP, and a rise in glutamate.

   e **True.** Reperfusion introduces excess oxygen radicals which interfere with the removal of free cell calcium ions.

19 a **False.** The sodium pump is found in the membranes of all mammalian cells.

   b **False.** The sodium pump is a protein that extends right through the membrane (a transmembrane protein).

   c **False.** Three Na^+ ions exit cell in exchange for entry of two K^+ ions.

   d **False.** Despite the expulsion of $3Na^+$ for $2K^+$, the net electrogenic effect of the pump is only around 4 mV.

   e **True.** The pump creates an electric current due to the unequal numbers of Na^+ and K^+ transported in each direction.

20 a **True.** The pump concentration in the cells specialized in ion transport (brain, kidney, intestine, and muscle) is over 100 times greater than in other cells.

   b **False.** Thyroid, catecholamines, insulin, and aldosterone stimulate the sodium pump in decreasing order of efficacy.

   c **True.** However, the direction of ion flux *in vivo* is fixed.

   d **True.** The pump also displays saturation kinetics.

   e **True.** Intracellular Mg^{2+} ions are necessary for ATPase activity.

21 a **True.** They can only attach to the extracellular part of the pump.

   b **False.** Both metabolic inhibitors block ATP production and so remove the pump's energy source.

   c **True.** Hepatic metabolism contributes a similar amount of heat.

   d **True.** It also helps maintain osmotic balance and protects the fragile cell membrane from rupture.

   e **True.** Other primary cation pumps have been identified: H^+, K^+-ATPase, Ca^{2+}, H^+-ATPase and H^+-ATPase. A Mg^{2+}-ATPase probably exists as well.

22 a **True.** The resting membrane has a very low permeability to sodium ions.

   b **False.** Other ions also play a part, e.g. Cl^- and HCO_3^-, but permeability to K^+ is the major determinant of resting membrane potential.

   c **True.** The membrane potential is produced by only a small percentage of the total ions which are present at the cell membrane boundary.

d **False.** Indiffusible ICF anions, mainly proteinate, hold cations such as K^+ inside the cell.

e **False.** Repolarization is caused by the exit of cations from the cell. Potassium is almost solely responsible. It flows out via voltage-gated K^+ channels down a steep electrochemical gradient.

23 a **False.** They are aqueous protein pores that traverse membrane proteins.

b **False.** Normal ionophores respond to ligands (L-gated), voltage change (V-gated), and intracellular calcium ions.

c **True.** Depolarization initiates the inward flow of sodium by rapidly opening **m** gates (1 ms); it also initiates slower closing of **h** gates which cause an inactive refractory state (several ms).

d **True.** Before the channel returns to a closed state which can be opened again, the cell membrane must be hyperpolarized by the opening of K^+ channels.

e **True.** The K^+ channel has a large number of subtypes. The Na^+ channel evolved from the Ca^{2+} channel in the Cambrian period.

pH, buffers, and pH abnormalities (Qs 24–34)

24 Regarding pH:

a Pure water with a pH of 7.0 contains no free hydrogen ions.

b With a plasma pH of 7.4 the red blood cell has a pH of about 7.20.

c The body is a net producer of acids under normal resting conditions.

d Pure gastric juice has a $[H^+]$ concentration of 0.1 M.

e Plasma pH is normally the same as the pK_a of carbonic acid in plasma.

25 The following statements about the pH of urine are true:

a The pH of urine is usually acid.

b Aciduria occurs shortly after eating a large protein meal.

c Alkaluria is a consistent finding in metabolic alkalosis (assume kidneys are normal).

d The H_2PO_4 of urine is known as titratable acid.

e Urine can contain considerable amounts of free acid.

26 Buffers:

a A buffer is best *in vivo* when half is present as weak acid, half as salt.

b $NaHCO_3/H_2CO_3$ is an important buffer system in both plasma and urine.

c The phosphate system is a better urinary buffer than the $NaHCO_3$ system.

d Haemoglobin (Hb) is of little importance as a blood buffer.

e Deoxygenated haemoglobin (Hb) is a better buffer of H^+ ion than oxy-haemoglobin (HbO).

27 *With regard to the ammonium buffer system (NH_4^+ = $NH_3 + H^+$):*

a   The ammonium buffer system has a pK_a of 9.3.
b   In the kidney NH_3^+ is mainly made and secreted in the distal tubules.
c   Over 90% of renal ammonia is derived from glutamine.
d NH_3 is lipid insoluble thus cannot diffuse back into luminal cells.
e   The ammonium ion (NH_4^+) is highly polar and crosses membranes poorly.

28 *A respiratory acidosis:*

a   is caused by a failure of the lungs to eliminate CO_2 adequately.
b   is associated with a fall in plasma bicarbonate.
c   is associated with hyperventilation in normal people.
d   is indexed well by measuring the standard bicarbonate.
e   is less likely than alkalosis in an acute asthmatic attack.

29 *The buffer base (BB):*

a   is the same as the base excess (BE) in a normal person.
b   is critically dependent on Hb concentration.
c   is the sum of the buffer cations in the plasma.
d   is expressed as mmol/l.
e   is negative in acidosis.

30 *Lactate in plasma:*

a   is not found in the blood of normal people at rest.
b   can produce an acidosis of 20–30 mmol/l in severe exercise.
c   even when oxygen is present, can only be metabolized by the liver.
d   causing a persistent acidosis is a major acid/base disturbance.
e   causing acidosis is a well-recognized complication of diabetes mellitus.

31 *Metabolic acidosis:*

a   is typical of chronic renal failure.
b   causes hyperventilation by stimulating the central chemoreceptors.
c   can be caused by hyperchloraemia.
d   can be caused by hyperkalaemia.
e   can cause hyperkalaemia.

32 *Development of a metabolic acidosis:*

a   is an early sign of salicylate (aspirin) intoxication.
b   is a consequence of ingesting ammonium chloride (NH_4Cl).
c   occurs in persistent vomiting from obstruction of the small bowel.
d   can be roughly quantified by knowing base excess and body surface area.
e   usually develops rapidly unless treatment is instituted promptly.

33 *The initial pH disturbance:*

a   in metabolic acidosis is a rise in lactic acid.

b   in metabolic alkalosis is an increase in urinary acid loss.

c   in respiratory acidosis is a rise in plasma bicarbonate.

d   in respiratory alkalosis is a decrease in plasma P_{CO_2}.

e   occurs as a change in the pH of the ECF.

34 *Blood gases:*

a   Arterial samples are superior to venous ones when assesing cardiopulmonary function.

b   Venous samples are more informative than arterial samples about pH status after a cardiac arrest.

c   The standard protocol dose of $NaHCO_3$ is safe in respiratory acidosis.

d   In clinical practice the 'blood gases' are taken to include plasma P_{O_2}, P_{CO_2}, pH, base excess, and $NaHCO_3$.

e   The blood gases constitute the most useful single test of acid–base status.

pH, buffers, and pH abnormalities (Answers)

24   a   **False.** At 25 °C pure H_2O contains 10^{-7} g of H^+/litre.

b   **True.** A lower pH is also found in resting muscle cells (pH 7.0) and in the CSF (pH 7.32). This is probably due to carbonic anhydrase in the CSF.

c   **True.** About 10 100 mmol of H^+ is produced per day (H_2CO_3 to CO_2 + H_2O) of which 10 000 mmol is volatile.

d   **True.** This means that pure acid gastric juice has a pH of 1.0.

e   **False.** This is only true when the acid is half dissociated (Henderson–Hasselbalch equation; pH = pK_a + log[salt]/[acid]).

25   a   **True.** Aciduria is the rule because of acids produced in metabolism and from our mostly acidic diet.

b   **False.** The postprandial 'alkaline tide' occurs independently of the pH of the food ingested.

c   **False.** Alkaluria occurs in acute alkalosis, aciduria is found in chronic cases.

d   **True.** This is because $H_2PO_4^-$ (and not NH_4^+) liberates H^+ if urine is titrated back to the pH of plasma.

e   **False.** Furthermore the distal tubule cannot meet the body's need to get rid of H^+ in this way. Most acid in the urine is in a buffered form.

26   a   **False.** True *in vitro*; in the body most of the buffer should be present as a salt since so much acid is being formed.

b   **False.** Normally urinary $NaHCO_3$ is very low (2 mmol/l).

c   **True.** Significant amounts of phosphate are found in urine especially in the distal tubules. Furthermore, the pK of phosphoric acid (6.8) is closer to the pH of the glomerular filtrate (7.4) than the pK of carbonic acid (6.1).

d **False.** Hb is an excellent physiological blood buffer.

e **True.** Deoxygenated Hb is a relatively weaker acid than oxygenated Hb.

27 a **True.** In any body fluid almost all NH_3 is present as NH_4 ions.

b **False.** Most NH_3 is formed and secreted in the proximal tubules. Most of the NH_3 secreted distally is actually formed proximally.

c **False.** About 60% of renal ammonia is derived from glutamine; 40% is made from glycine and alanine.

d **False.** NH_3 is lipid soluble and freely diffusible.

e **True.** The failure of NH_4^+ to cross from lumen to cell is the basis for the urine excretion of NH_3-buffered H^+.

28 a **True.** There are three reasons for this: disturbed perfusion, ventilation, or muscular impairment.

b **False.** The retention of CO_2 causes a rise in plasma bicarbonate.

c **False.** Hyperventilation normally causes respiratory alkalosis.

d **False.** Standard bicarbonate is altered in metabolic pH disturbances.

e **True.** In acute asthma arterial blood typically shows hypoxaemia, increased Pco_2 and a fall in pH, viz. a respiratory acidosis.

29 a **False.** BE = observed BB (OBB) minus normal buffer base (NBB).

b **True.** The buffer base line is constructed using bloods of differing Hb concentration.

c **False.** It is the sum of the buffer anions of blood (HHb^- + HCO_3^-).

d **True.** It is normally 48 mmol/l (Hb = 21; HCO_3 = 27).

e **True.** BE = OBB − NBB; an example in acidosis might be −18 = 30 − 48.

30 a **False.** Resting venous blood contains about 1 mmol lactate/l.

b **True.** Higher lactate levels will cause severe muscle fatigue.

c **False.** About 25% of lactate is oxidized to CO_2 and H_2O in other body tissues.

d **True.** Lactic acidosis is biochemically defined as a base deficit > 6.0 mmol/l; a pH < 7.35, and an arterial lactate > 5 mmol/l.

e **True.** Lactic acidosis also occurs in shock, renal disease, liver disease, with alcohol excess, in familial and idiopathic lactic acidosis, and with drugs, e.g. phenformin.

31 a **True.** The kidneys have a major role in the maintenance of pH by their excretion of fixed acids.

b **False.** Hyperventilation is due to stimulation of carotid bodies.

c **True.** Excess chloride upsets the $NaHCO_3/H_2CO_3$ equilibrium by removing some of the sodium from bicarbonate, thus producing an acidosis.

d **True.** Potassium enters cells and displaces hydrogen ion into ECF.

e **True.** Excess hydrogen ions enter cells and displace potassium.

32 a **False.** Initial respiratory alkalosis (overcompensation) occurs.

  b **True.** NH_3 is incorporated into urea in the liver, the inability to completely buffer any HCl that is produced results in an acidosis.

  c **True.** Acidosis in small bowel vomiting is chiefly due to ketogenesis and partly due to a loss of gut alkali.

  d **False.** $0.3 \times$ base excess $\times$ body wt (kg) = mmol bicarbonate needed.

  e **False.** Compensation (pulmonary and renal) slows the progress of acidosis. Pulmonary compensation is more rapid but a long-standing acidosis will be compensated for by HCO_3 retention by the kidneys.

33 a **False.** Primary abnormality is a fall in plasma bicarbonate.

  b **False.** Primary abnormality is a rise in plasma bicarbonate.

  c **False.** Primary abnormality is a rise in plasma Pco_2.

  d **True.** Bicarbonate changes are secondary.

  e **True.** ICF is not readily accessible for pH examination.

34 a **True.** Although relatively constant a–v differences make venous sampling satisfactory if one is only interested in acid–base balance.

  b **True.** Venous sampling gives more informative results after cardiac arrest (where there is an inability of lungs to excrete CO_2).

  c **False.** Bicarbonate may interact with lactic acid and generate high venous Pco_2 which is retained if pulmonary failure exists.

  d **True.** Although technically nitrogen is also a blood gas.

  e **True.** However, beware of compensation and of overcompensation such as occurs in salicylate intoxication (Q32).

Hypovolaemia (Qs 35–41)

35 *Sweating:*

  a can account for over 3 litres of fluid loss per hour.

  b occurs continually at a low rate even in temperate climates.

  c at low rates of flow may mean that the Na^+ in sweat falls as low as 1 mmol/l.

  d at high rates of flow allows large losses of sodium to occur.

  e cannot occur when there is cutaneous vasoconstriction.

36 *About parenteral infusions:*

  a Ringer's solution contains sodium, chloride, potassium, and calcium.

  b Hartman's solution is another name for Ringer's solution.

  c Most parenteral infusions are well absorbed by the intraperitoneal (IP) route.

  d Darrow's solution is ideal when treating fluid loss in hypokalaemia.

  e One-sixth molar solutions are hypotonic.

37 When vomiting occurs:

  a Gastric vomiting causes an alkalosis.
  b Vomitus consisting of antral gastric secretions is alkaline.
  c Nasogastric suction is associated with acidosis.
  d pH of vomitus is approximately neutral in small bowel obstruction.
  e Reverse peristalsis is typical.

38 Vomiting:

  a is a true reflex action.
  b is controlled by the vomiting centre in the hypothalamus.
  c is associated with the area postrema where the blood–brain barrier is absent.
  d is associated with glottis closure.
  e is physically caused by contraction of skeletal muscle.

39 Thirst:

  a is sensed by a thirst centre found in the brain stem.
  b is a sensitive monitor of water balance.
  c is stimulated by increased plasma osmolality.
  d is stimulated by a dry pharynx.
  e only disappears when isovolaemia is restored

40 In altered bowel function:

  a Most cases of diarrhea are caused by a disturbance of function in large bowel.
  b Hypernatraemic dehydration can occur in infants with diarrhea.
  c Autonomic neuropathy, e.g. in diabetes, often causes constipation.
  d Many agents cause diarrhea by stimulating chloride secretion.
  e Diarrhea may cause metabolic acidosis or hypokalaemia, depending on the segment of the bowel involved.

41 Oral rehydration solution (ORS):

  a Standard isotonic ORS reduces the rate of stool loss in diarrhea.
  b Glucose-based ORS is an important source of calories.
  c In ORS a carbohydrate : sodium ratio of about 1 is ideal for maximal H_2O and electrolyte absorption.
  d Cooked cereal powders, e.g. rice, may be substituted for glucose.
  e The tonicity of the ORS is critical in its proper use.

Hypovolaemia (Answers)

35 a **False.** Maximal sweating recorded is just over 1.5 litres/h.
    b **False.** Sweating occurs in response to hyperthermia or emotion.
    c **False.** At low rates of flow, sweat Na^+ may be as low as 5 mmol/l.

    d **True.** In profuse sweating, 10 litres of sweat contains 23 g sodium.

    e **False.** Note there is a 'cold sweat' of fear.

36 a **True.** 1 litre of Ringer's contains Na$^+$ 146 mmol, Cl$^-$ 152 mmol, K$^+$ 4 mmol, and Ca^{2+} 2 mmol.

    b **False.** Hartman's is Ringer's except that 28 mmol Cl$^-$ is replaced by lactate.

    c **True.** The intraperitoneal route is often an excellent alternative to the intravenous route. However, it is rarely used clinically because of fear of complications.

    d **True.** Darrow's contains 35 mmol/l of potassium.

    e **False.** One-sixth molar solutions are isotonic. ECF is 1/6 molar (1/3 osmolar).

37 a **True.** Pure parietal juice is 9 parts HCl, 1 part KCl (0.5 M HCl).

    b **True.** Antral juice is rich in NaCl and NaHCO$_3$; 1 ml/h is secreted at rest.

    c **False.** Nasogastric suction removes acidic gastric secretions and therefore is associated with an alkalosis.

    d **True.** Alkaline small bowel secretions are mixed with acid gastric ones.

    e **False.** A reverse contraction of small bowel may precede vomiting. Reverse peristalsis occurs regularly in retching.

38 a **True.** Vomiting is a complex co-ordinated reflex like swallowing or coughing.

    b **False.** The vomiting centre is in the area postrema of the brain stem.

    c **True.** Thus drugs such as apomorphine or digoxin are emetics since they cross the blood–brain barrier at this point.

    d **True.** Closure of the glottis prevents aspiration of vomitus.

    e **True.** Co-ordinated contraction of abdominal muscles and diaphragm are the immediate causes of vomiting.

39 a **False.** The thirst centre is in the hypothalamus just posterior to the optic chiasma.

    b **False.** Thirst is a crude emergency signalling system. None the less our weight only alters by 0.5% per day.

    c **True.** Both hypertonic ECF and hypovolaemia are powerful stimulants of thirst.

    d **True.** A dry pharynx stimulates thrist independently of plasma tonicity but, unlike true thirst, it is diminished by simply wetting the oral cavity.

    e **False.** Even a few mouthfuls of water can slake thirst. Note that atrial natriuretic factor (ANF) is an antithirst factor. Evidence exists for volume receptors in the oesophagus allowing assessment of fluid input before changes in osmolality occur.

40 a **False.** Most cases of diarrhea are due to disease of the small intestine.

    b **True.** Large bowel reabsorbs Na$^+$ but loses ability to reabsorb H$_2$O.

    c **False.** Motility defect leads to bacterial overgrowth and diarrhea. Constipation rarely occurs.

d **True.** Examples include cholera, *E. coli*, and vasoactive intestinal polypeptide (VIP).

e **True.** Acidosis occurs from a loss of $NaHCO_3$ in small bowel diarrhea and hypokalaemia from loss of potassium in large bowel diarrhea.

41 a **False.** Nor does glucose/electrolyte ORS limit the duration of illness. Note that hypotonic ORS (224 mosm/kg) appears to reduce stool output by 28% compared with standard solutions.

b **False.** The recommended ORS contains 20.0 g glucose/l, viz. 80 kcal/l.

c **True.** However, a ratio of up to 3:1 has proved successful.

d **True.** Cereal-based ORS (50–80 g cooked rice powder) is more effective in co-transporting Na^+ than glucose.

e **True.** An ORS that is hypertonic aggravates existing diarrhea.

2

Haematology

BARTS & THE LONDON QMSMD

Erythropoietin and haemoglobin (Qs 1–9)

1 Erythropoietin:

a is synthesized by cells in the afferent arteriole in the kidney.
b release is stimulated by a rise in plasma CO_2.
c is a mucopolysaccharide.
d can also be made by the liver.
e is a cytokine.

2 Clinically, erythropoietin:

a is ony effective if obtained from human cadavers.
b is often useful in chronic renal failure (CRF).
c may be valuable in certain cases of severe pruritus.
d may be used in treating sickle-cell disease.
e affects erythropoiesis primarily by increasing the production of reticulo-cytes.

3 Haemoglobin:

a contains six iron atoms.
b can carry six oxygen atoms.
c contains iron which is primarily centred in a tetrapyrrol ring.
d is too large to pass through the renal glomerular membrane.
e is synthesized by nucleated red cells in the bone marrow.

4 The form of haemoglobin:

a All Hb should be in the form of HbA in red cells of a normal adult.
b HbA_2 has two alpha- and two delta-chains.
c Sickle-cell haemoglobin (HbS) has a substitution in its alpha-chain.
d Hb procured by replacement of the sixth beta-chain glutamic acid by lysine is only mildly functionally impaired.
e HbH (tetramer of four beta-chains) is found in beta-thalassaemia.

5 Haemoglobin F:

a is only found in the fetus.
b has a high P_{50}.
c has a similar oxygen dissociation curve to that of myoglobin.
d increases after erythropoietin therapy.
e contains gamma- and beta-chains that differ from each other by fewer than five amino acids.

6 In sickle-cell disease:

a HbS confers some protection against *P. falciparum*.
b There is a similar prognosis in HbAS (sickle-cell trait), HbSC (SC disease), and HbAC (sickle C disease).

c Raised HbF levels are protective against the clinical symptoms of SS disease.

d The solubility of HbS falls to about 50% of normal in hypoxia.

e Obstruction of the microcirculation is typical.

7 Methaemoglobin (MetHb):

a contains Hb in which iron is replaced by copper.

b has an oxygen dissociation curve similar to that of HbF.

c has a constant tendency to form *in vivo*.

d is one cause of a 'blue baby'.

e may be caused by an excess intake of spinach.

8 Glycosylated Hb:

a is the combination of a HbA and a sugar, e.g. HbA1c which is a combination of HbA1 and a sugar.

b and glycosylated plasma proteins are a normal finding in blood.

c measured as HbA1c in plasma gives a more accurate retrospective estimate of blood sugar levels than other gylcosylated products.

d HbA1c levels are a good index of glucose-induced arteriolar dilation and consequent damage to renal function in diabetics.

e plus other advanced glycosylation end products (AGEs) are highest in infants.

9 Carboxyhaemoglobin (COHb):

a can cause a severe cyanosis.

b is normally found in human blood.

c is increased in cigarette smokers.

d is increased in severe exercise.

e is about 200 times more likely to form than oxyHb with the exposure of Hb to equal partial pressures of CO and O_2.

Erythropoietin and haemoglobin (Answers)

1 a **False.** Erythropoietin is synthesized by the epithelial cells of peritubular capillaries.

b **False.** Erythropoietin release is stimulated by hypoxaemia.

c **False.** Erythropoietin is a glycoprotein.

d **True.** The liver is the chief source of erythropoietin in young infants.

e **True.** Erythropoietin is a growth factor like interleukins (ILs) and colony-stimulating factors (CSFs).

2 a **False.** Recombinant erythropoietin is available but it is expensive.

b **True.** Erythropoietin is useful for treating the refractory anaemia of CRF.

c **True.** Erytyhropoietin is especially useful in uraemic pruritus.

   d **True.** Erythropoietin inhibits polymerization of deoxyHbS and increases the number of gamma-chains, thus fewer abnormal beta-chains remain.

   e **False.** Erythropoietin stimulates all red cell progenitors and helps to maturate as well to generate reticulocytes.

3   a **False.** Each Hb molecule contains four iron atoms.

   b **True.** When fully oxygenated a Hb molecule can carry four O_2 molecules (eight atoms).

   c **True.** Iron is centred in a tetrapyrrol ring.

   d **False.** Haemoglobinuria often occurs in haemolytic states.

   e **True.** Hb synthesis occurs throughout the normal 6–8 days of erythroid cell maturation.

4   a **False.** Normal adult Hb is a mixture of 97% HbA; 2.0% HbA_2; 1.0% of other variants.

   b **True.** HbA_1 = two alpha- and two beta-chains; HbA_2 = two alpha- and two delta-chains; HbF = two alpha- and two gamma-chains.

   c **False.** In HbS, valine replaces glutamine in the sixth position along the beta-chain.

   d **True.** This is HbC. Only mild sickling occurs even in homozygous cases. HbC is found in 3% of Afro-Americans.

   e **False.** HbH signifies alpha-thalassaemia. The subject cannot form alpha-chains.

5   a **False.** Neonatal Hb is chiefly HbF. Note, too, that adult Hb is about 0.5% HbF.

   b **False.** HbF has a low P_{50} (see oxygen dissociation curve).

   c **True.** The curves for myoglobin and HbF are both placed well to the left of that for HbA.

   d **True.** Young, immature red cells have a relatively high concentration of HbF.

   e **False.** Gamma- and beta-chains differ by 39 amino acid residues out of 146.

6   a **True.** Red cells containing HbS are unsuitable for the development of *P. falciparum*.

   b **True.** All three contain roughly 20–50% HbS. Note the particular tendency to thrombosis and embolism in pregnancy in SC disease.

   c **True.** HbF directly inhibits polymerization of HbS by forming mixed hybrids ($alpha_2$–beta–gamma) with much higher solubility than HbS. The high levels of HbF in some communities in eastern Saudia and southern India (where 20–40% of Hb in adults is HbF) gives considerable protection against sickling. Note: the cytostatic agent hydroxyurea works by raising HbF and by producing Hb $alpha_2$-beta-gamma.

   d **False.** Solubility of HbS falls to about 2% of normal in hypoxic conditions.

   e **True.** Sickle cells aggregate, haemolyse, and obstruct vessels.

7   a   **False.** In MetHb iron is in the oxidized ferric form (Fe^{3+}).

    b   **False.** MetHb cannot carry oxygen.

    c   **True.** The tendency to form MetHb is countered by plasma methaemoglobin reductase.

    d   **True.** The bluish discoloration caused by MetHb is easily mistaken for cyanosis.

    e   **True.** Any food rich in nitrates and nitrites can cause methaemoglobinaemia.

8   a   **True.** Glycosylation of Hb involves the non-enzymatic binding of a hexose to the N-terminal amino acid of the beta-chain.

    b   **True.** Glycosylation even affects proteins of the vessel wall.

    c   **False.** Glycosylated serum proteins, e.g.fructosamine, are more accurate, especially in reflecting plasma glucose levels of the preceding 1–3weeks.

    d   **True.** The risk of microalbuminuria increases logarithmically when HbA1c exceeds 8.1% (corresponding to an average daily blood glucose of 200 mg or 11.1 mmol/l).

    e   **False.** AGEs are low in infants, they increase in ageing, with diabetic complications and in atheroma.

9   a   **False.** COHb typically has a typical cherry-pink colour. However, in a recent series of 100 cases only one had this discoloration.

    b   **True.** COHb is always present in small amounts in red cells.

    c   **True.** COHb comprises up to 10% of Hb in heavy smokers.

    d   **False.** Exercise and the accompanying hyperventilation lower COHb.

    e   **True.** Thus the rapidity of CO poisoning. COHb cannot carry oxygen; CO is also toxic to cytochromes. Note that NO also combines with the iron in Hb and cytochromes, as well as with the iron in guanylate cyclase.

Red cell metabolism, iron, and anaemia (Qs 10–27)

10 *In red blood cell metabolism:*

    a   no ATP is formed.

    b   2,3-diphosphoglycerate (2,3DPG) rises to approximately equimolar amounts with Hb.

    c   2,3DPG raises the pH.

    d   a special cytochrome system is involved.

    e   short-chain fatty acids are readily utilized.

11 *The red blood cell:*

    a   life span can be determined by the Kleihauer test.

    b   requires the presence of insulin for the easy entry of glucose.

c integrity depends largely on the presence of reduced glutathione (GSH).
d contains carbonic anhydrase (CA) which is structurally identical to that found in the kidney.
e shows significant deterioration in its ability to carry oxygen with ageing.

12 The spleen:

a has a sparse lymphatic supply.
b can capture and process circulating antigen.
c has a major haematopoietic function in the fetus and newborn.
d is the most probable site for the production of tuftsin.
e if removed, is not followed by any important long-term adverse effects.

13 Regarding functions of the spleen:

a It is a reservoir for blood platelets.
b Pitting refers to the ability of the spleen to extract cytoplasmic inclusions and foreign material from RBCs, or even to remove nuclei from normoblasts.
c The spleen is a more important immunological organ in children than in adults.
d Howell–Jolly bodies are concretions of iron in erythrocytes and are often seen in a normal spleen.
e Following splenectomy, thrombocytosis lasts for months or years.

14 Reticulocytes:

a have no nucleus.
b are unable to synthesize haemoglobin.
c contain mitochondria.
d have surface receptors for plasma transferrin.
e normally comprise 5–6% of circulating red blood cells.

15 The enzyme glucose-6-phosphate dehydrogenase (G6PD):

a This is the only means for the RBC to generate NADPH.
b When this enzyme is lacking Heinz bodies may appear in erythrocytes.
c G6PD deficiency is transmitted as an autosomal recessive gene.
d In G6PD deficiency aspirin and acetaminophen may induce haemolysis.
e Female heterozygotes for G6PD are more resistant to falciparum malaria than normal males or females.

16 Iron:

a Intake/day should at least equal iron used daily in erythropoiesis.
b Women have greater stores of iron, per kg body wt, than men.
c has a high bioavailability from cast-iron cooking utensils.
d absorption occurs mostly in the jejunum.
e There is preferential absorption of iron from meat and meat products compared to that from vegetables.

17 *In iron metabolism:*

a absorption is regulated by the ferritin status of the mucosal cell.

b mucosal ferritin can become apoferritin if excess iron is taken.

c mucosal ferritin enters the blood and is carried to the liver.

d plasma transferrin carries iron to body stores.

e iron may constitute 50% of the weight of haemosiderin.

18 *Iron overload:*

a involves the formation of ferritin by most body cells, including erythrocytes.

b when complicated by a deposition of melanin in the skin and beta cells of the pancreas can cause 'bronzed diabetes'.

c is typical of beta-thalassaemias.

d treatment uses Fe chelation, avoiding Fe in diet, and performing repeated phlebotomies.

e in haemochromatosis is primarily caused by hereditary factors in those of African descent.

19 *Vitamin B_{12} (methylcobalamin (MCB) and deoxyadenosylcobalamin (DOCB)):*

a causes a megaloblastic macrocytic anaemia.

b causes a rise in cell methyltetrahydrofolate (MTHF).

c causes a rise in cell tetrahydrofolate (THF).

d is carried by transcortin in the plasma.

e is necessary for the formation of purines and pyrimidines.

20 *Vitamin B_{12}:*

a is present in foods of animal origin only.

b stores in the liver can provide body needs for up to 2 years.

c deficiency significantly affects white blood cells.

d decrease in pernicious anaemia patients is mostly due to antibodies to intrinsic factor (IF).

e Treatment of pernicious anaemia with B_{12} may worsen neurological symptoms.

21 *Folic acid:*

a deficiency may be due to fish tapeworm disease, particularly in Scandinavia.

b is a fat-soluble vitamin.

c is essential for normal development and closure of the neural tube.

d potentiates the effect of phenytoin in epilepsy.

e deficiency is usually associated with worse GIT symptoms than those in patients with B_{12} deficiency.

22 *Regarding red blood cell turnover:*

a The rate of destruction of red cells is partly controlled by an endocrine mechanism.

b Reticulocytes are less vulnerable to destruction than erythrocytes.
c Fetal and neonatal red cells live about 120 days.
d The spleen is normally the main organ of RBC destruction.
e Red cell breakdown can occur in the skin of normal people.

23 In haemoglobin (Hb) turnover:

a Haemoglobin is not normally released intravascularly.
b When there is intravascular haemolysis, haemopexin combines with free haem in the plasma.
c Bilirubin is a stable derivative of biliverdin.
d Plasma bilirubin is derived exclusively from haemoglobin.
e Bilirubin that is released by tissue macrophages is water soluble.

24 Bilirubin:

a Dicoumarols compete with bilirubin for plasma albumin.
b Prehepatic bilirubin gives a positive indirect van den Bergh reaction.
c Prehepatic bilirubin rises in haemolytic jaundice.
d Sialic acid is the acceptor binding compound for bilirubin at the liver cell.
e Gilbert's disease is the most common cause of familial haemolytic jaundice.

25 In obstructive jaundice:

a itch is a common problem.
b a palpable gall bladder suggests a non-gall bladder cause, e.g. cancer of the head of the pancreas.
c kernicterus is a threat, especially in infants.
d occurrence may have been caused by severe stricture or a large stone in the cystic duct.
e the urine has a dark reddish colour.

26 Haemolytic anaemia:

a is often caused by immune haemolysins.
b may give rise to a direct positive Coombs test.
c is the most common cause of anaemia in children.
d with anticomplement antibodies is suggestive of paroxysmal nocturnal haemoglobinuria (PNH).
e can be due to spherocytes which use more energy than normal erythrocytes.

27 Plasma proteins:

a are approximately 10 g/dl in Western societies.
b contain immunoglobulins which are larger than fibrinogen.
c provide the colloid osmotic pressure of plasma.
d normally constitute the major contributor to plasma viscosity.
e are virtually all formed in the liver.

Red cell metabolism, iron, and anaemia (Answers)

10 a **False.** ATP is formed during glycolysis in the red cell.

    b **True.** About 5 mM of each. Very little 2,3DPG in other tissues.

    c **False.** 2,3DPG lowers ICF pH, thus facilitating deoxygenation of HbO_2.

    d **False.** The mature red cell has no mitochondria and therefore no cytochromes.

    e **False.** The red cell relies solely on the metabolism of glucose.

11 a **False.** The Kleihauer test identifies fetal Hb by electrophoresis.

    b **False.** Insulin is not needed for glucose entry into the red cell, neural tissue, renal tubules, GIT, hepatocytes, placenta.

    c **True.** GSH protects against damaging oxidants.

    d **False.** Several variants of CA have now been identified.

    e **False.** Fortunately 'old' red cells carry oxygen just as well as 'young' ones.

12 a **False.** The spleen is largely composed of lymphoid tissue.

    b **True.** An example is Ig-tagged salmonella bacteria which are trapped by splenic macrophages.

    c **True.** Reversion to haematopoiesis in the spleen in an adult occurs only in disease.

    d **True.** Tuftsin stimulates phagocytosis.Tuftsin deficiency occurs in AIDS. Note controversy exists about the role of tuftsin in the body .

    e **False.** There is grossly impaired resistance to malaria, bartonella, and pneumococcal infections in postsplenectomy patients. Children are put on prophylactic penicillin and all splenectomy patients should be vaccinated against pneumococcus.

13 a **True.** Platelets spend 3 days (one-third of their life span) in the spleen.

    b **False.** Pitting refers to the removal of remnant nuclear DNA from young erythrocytes.

    c **True.** None the less one should give adults antipneumoccocal vaccine after a splenectomy.

    d **False.** H–J bodies are DNA remnants. Typical of RBCs of those with a non-functional spleen, e.g. autoinfarction in SS disease, or in post-splenectomy states.

    e **True.** Post-splenectomy thrombocytosis rarely requires treatment or poses problems.

14 a **True.** A reticulocyte is formed by the extrusion of the nucleus from a normoblast.

    b **False.** Reticulocytes synthesize Hb at low levels for 1–2 days.

    c **True.** Reticulocytes contain residual cytoplasmic RNA and mitochondria.

    d **True.** In this way reticulocytes can trap iron for Hb synthesis.

    e **False.** Normal reticulocyte range is 0.5–1.5%. A reticulocytosis indicates an increased rate of erythrocyte production, e.g. in haemolytic anaemia.

15 a **True.** Absence of NADPH in the red cell renders the cell liable to oxidative damage.

b **True.** Heinz bodies are concretions of MetHb and glutathione.

c **False.** G6PD deficiency is transmitted as an X-linked recessive gene.

d **False.** Haemolytic agents in G6PD deficiency include dapsone, nalidixic acid, primaquin, fava beans, and infections.

e **True.** No satisfactory explanation for the relative resistance against malaria shown by female heterozygotes for G6PD deficiency has been found.

16 a **False.** Most iron for new RBCs is recycled from catabolized Hb.

b **False.** Males store about 500 mg Fe; females store about half this.

c **True.** A meal of chicken and vegetable balti curry can provide about 50 mg of dietary iron from the cast-iron container. Note total body iron is 3–5 g of which 2.5–3.5 g is in Hb.

d **True.** Iron is mostly absorbed from the upper small gut. Only a trace is absorbed in the stomach. None the less a relatively low pH promotes solubility of the Fe^{2+} ion.

e **True.** 20–40% haem iron is absorbed; only 5% of vegetable iron is absorbed.

17 a **True.** Mucosal cells that are full of ferritin cannot accept more iron.

b **False.** The opposite is the case. Apoferritin = ferritin minus iron.

c **False.** Mucosal ferritin passes Fe^{2+} to plasma transferrin. This is all lost in the faeces if the mucosal cell is shed into the lumen of the gut before transfer occurs.

d **True.** Iron stores are in the liver, the bone marrow, and the spleen.

e **True.** Ferritin molecules degrade and coalesce to form the macromolecule haemosiderin.

18 a **False.** Only nucleated cells can form ferritin. Thus Fe can be deposited almost anywhere, e.g. in the heart, endocrine organs, viscera, and skin.

b **False.** Haemosiderin (not melanin) under the skin and in the beta cells of pancreas gives rise to 'bronzed diabetes'.

c **True.** Iron overload can easily occur in thalassaemia because of the excess iron absorption from repeated blood transfusions.

d **True.** Iron chelators include desferrioxamine and the newer specific α-ketohydroxypyridines.

e **False.** In haemochromatosis genetic factors are not nearly so important in Africans as in those of northern European descent.

19 a **False.** A lack of vitamin B_{12} causes anaemia.

b **False.** Normally MTHF donates methyl groups to form active B_{12} (MCB).

c **True.** This occurs because MTHF no longer forms THF.

d **False.** Vitamin B_{12} is carried by transcobalamin. Transcortin carries cortisol.

e **True.** Purines and pyrimidines are the constituents of DNA.

20   a   **True.** True vegans may develop vitamin B_{12} deficiency.

    b   **False.** Up to 3 mg of B_{12} is stored in the liver. This should last about 4–5 years.

    c   **True.** B_{12} is needed for DNA synthesis. A deficiency of B_{12} is clearly manifested in nucleated and especially in dividing cells.

    d   **False.** Anti-IF is found in 40–50% of cases of pernicious anaemia that are due to autoimmunity.

    e   **True.** Hypothesis: increased RBC thiocyanate oxidase converts thiocyanate to neurotoxic cyanide.

21   a   **False.** *Diphyllobothrium latum* can cause vitamin B_{12} deficiency.

    b   **False.** Folate is water soluble. It is found especially in leafy vegetables.

    c   **True.** Link between folate and neural tube development was first mooted in the 1960s and was proved in 1992 by reports from Hungary. It is also likely that folate protects against cardiovascular system (CVS) disease. Folate, B_6, and B_{12} have a key role in homocysteine metabolism.

    d   **False.** Folate lowers plasma anticonvulsant level and vice versa.

    e   **True.** Cheilosis, glossitis, malabsorption, and diarrhea may be very severe in those with folate deficiency.

22   a   **False.** The intrinsic vulnerability of RBCs is the determining factor in red cell destruction.

    b   **False.** Reticulocytes are more vulnerable to destruction. They take 24 h to negotiate the splenic pulp.

    c   **False.** Fetal and neonatal red cells only live 45–70 days.

    d   **True.** None the less the liver and bone marrow easily cope with the job of destroying old, effete red cells after splenectomy.

    e   **True.** For example, this occurs in bruising (local macrophages are involved).

23   a   **False.** About 10% of daily breakdown of RBCs is intravascular.

    b   **True.** Haptaglobin is another plasma protein with a similar function to haemopexin.

    c   **True.** Biliverdin can be reduced to stable bilirubin.

    d   **False.** 10% of plasma bilirubin is derived from non-Hb porphyrins.

    e   **False.** Prehepatic bilirubin is fat soluble.

24   a   **True.** Salicylates and sulphonamides also compete for plasma albumin.

    b   **True.** Prehepatic bilirubin must be solubilized by adding alcohol to get the typical violet colour.

    c   **True.** Overt jaundice is not apparent until serum bilirubin > 3 mg%.

    d   **False.** Ligandin is the acceptor binding protein

    e   **False.** Gilbert's disease is the most common form of familial non-haemolytic jaundice (bilirubin usually < 100 μmol/l). It is due to a generalized defect of hepatic uptake of organic ions, including bilirubin. A decreased bilirubin UDP glucuronosyl-transferase type 1 is typical.

25 a **True.** The itch is probably due to retention of pruritic bile salts.

b **True.** Courvoisier's sign is the presence of a palpable gall bladder when jaundice is due to cancer of the pancreas or other obstruction of the biliary outflow.

c **False.** Kernicterus is due to prehepatic hyperbilirubinaemia. The infant's blood–brain barrier is poorly developed.

d **False.** The cystic duct joins the gall bladder to the common hepatic duct, forming the common bile duct, and so its obstruction cannot cause jaundice.

e **True.** Urine is loaded with readily filtered conjugated bilirubin in post-hepatic jaundice.

26 a **True.** Three types of antibody are found in haemolytic anaemias: direct exogenous antibodies, warm antibodies, and cold antibodies.

b **True.** Coombs serum has antibodies against human Ig and complement.

c **False.** Fe deficiency is the most common cause of anaemia in all age groups in the Western world.

d **True.** In PNH a membrane glycoprotein defect allows lysis by complement. Anticomplement antibodies are not diagnostic.

e **True.** Extra energy is used to expel Na^+, as spherocytes are more permeable to Na^+. A defect exists in the cytoskeleton proteins spectrin and ankyrin.

27 a **False.** The normal level of plasma proteins is approximately 7.0 g/dl (albumin 4.5 g/dl, globulin 2.5 g/dl).

b **False.** Fibrinogen (0.3 g/dl) is about 750 angstroms, gammaglobulin is about 250 angstroms.

c **True.** Colloid osmotic pressure (oncotic pressure of plasma proteins) is about 27 mmHg or 1.0 mosmol/kg.

d **True.** However, RBCs are the major contributor to whole blood viscosity.

e **False.** Note protein hormones, endothelial-derived proteins, non-hepatic alkaline phosphatase.

Blood groups and blood transfusions (Qs 28–34)

28 With ABO blood groups:

a The ABO factors are found on the surface of leucocytes.

b Most people secrete ABO factors into saliva, semen, and other bodily secretions.

c Group B non-secretors are very prone to urinary *E. coli* infections.

d Group A is associated with peptic ulcer.

e The ABO system is well established at birth.

29 In blood transfusion:

a A full cross match means mixing recipient's plasma with donor's RBCs.

b Most recorded deaths from blood transfusion are due to ABO incompatibility.

c About 90% of ABO incompatible transfusions are fatal (at least 1 unit transfused).

d Group O patients are more likely than those of other ABO groups to die from ABO incompatibility.

e ABO antibodies are present in all adults except those with group AB.

30 In allogenic blood transfusion:

a calcium ought to be given regularly during repeated transfusions.

b peritransfusion shivering is likely due to inorganic pyrogens.

c whole blood may be safely transfused intraperitoneally.

d it is safe to give O Rh-negative blood ('universal donor') to a bleeding patient whose group is unknown until she or he is normovolaemic.

e if no HIV antibodies are found in blood, the risk of HIV transmission by transfusing that blood is eliminated.

31 For blood transfusion:

a the accepted Hb threshold for perioperative transfusion has risen in recent years.

b a packed cell volume (PCV) of about 30% is well tolerated in surgery.

c autologous blood transfusion cannot be used in an emergency.

d the antiprotease drug, aprotonin, may lessen bleeding in surgery.

e hirudin, from the medicinal leech, is a good haemostatic agent

32 Concerning materno-fetal blood group incompatibility:

a Materno-fetal (M-F) ABO incompatibility is very rare.

b A large number of spontaneous abortions are due to M-F ABO mismatch.

c About 20% of newborns have mild ABO-induced haemolysis.

d M-F blood group incompatibility is more likely to cause problems if labour is prolonged and traumatic.

e Haemolytic consequences of M-F incompatibilty are likely to lessen with repeated pregnancies.

33 Rhesus (Rh) factor:

a is found in over 80% of the general population.

b is transmitted by a sex-linked dominant gene.

c antigen is confined to the red blood cell.

d negative people normally contain small amounts of anti-D in their plasma.

e positive cells are detected using the Kleihauer acid elution test.

34 Rh disease of the newborn:

a Always occurs where the mother is Rh-negative and the father is Rh-positive.

b Always occurs if the father is DD and the mother is dd.

c Feto-maternal transfusion occurs during a normal pregnancy.

d This cannot occur in a first pregnancy.

e A feto-maternal haemorrhage of greater than 15 ml is abnormal.

Blood groups and blood transfusions (Answers)

28 a **True.** None the less the greatest concentration of ABO factors is found on the surface of RBCs.

b **True.** 80% of people are 'secretors'.

c **True.** Particularly recurrent coliform urinary tract infection (UTI) in women who are group B non-secretors.

d **False.** Group O is associated with peptic ulcer. Group A is associated with cancer of the GIT (subjects express Forssman antigen).

e **False.** The ABO system is poorly established at birth.

29 a **False.** In a correct cross match one must also mix recipient's RBCs with donor's plasma.

b **True.** Most blood transfusion related deaths are caused by procedural errors resulting in ABO incompatibility.

c **False.** Only about 1 in 10 cases of ABO incompatible transfusion are fatal.

d **True.** Because group O patients have the highest level of anti-A and anti-B in the plasma.

e **True.** ABO antibodies develop after the first 3–6 months of life.

30 a **True.** Otherwise dangerous hypocalcaemia may develop with repeated transfusion.

b **False.** Peritransfusion shivering is likely due to allogenic complement and leucocyte interleukins (ILs).

c **True.** Intraperitoneal transfusion of small volumes of whole blood can be used in an emergency where facilities are poor. Is especially useful in infants.

d **False.** Rising level of α, β antibodies may damage patients RBCs.

e **False.** Time may elapse before the blood of a HIV-infected subject becomes HIV positive.

31 a **False.** The threshold has fallen so that nowadays transfusion is only justified if the Hb concentration is less than 7 g/l. It is now accepted that a Hb as low as 8 g/dl does not increase perioperative morbidity. The circulating blood volume and patient's oxygen requirements are as important as the Hb level.

b **True.** A low PCV increases blood fluidity and nutritional blood flow.

c **False.** Autologous blood is ideal if available, e.g. in ruptured ectopic pregnancy.

d **True.** Aprotonin is antikallikrein and antiplasmin. It has been found to be useful in cardiac surgery.

e **False.** Hirudin (antithrombin) is primarily an antithrombotic agent.

32 a **False.** M-F ABO incompatibility occurs in about 20% of all births.
   b **True.** M-F ABO incompatibilty is estimated to contribute to 30–40% of all natural abortions.
   c **False.** ABO-induced haemolysis probably occurs in about 3.0% of births. Neonatal red cells carry very little ABO antigen.
   d **True.** There is greater feto-maternal transfusion in traumatic or prolonged labour.
   e **False.** Haemolytic consequences likely to worsen with repeated pregnancies and repeated exposure of the mother to fetal antigen.

33 a **True.** About 86% of people are Rh-positive.
   b **False.** The Rh factor is transmitted as an autosomal dominant gene.
   c **True.** The Rh system is unlike the ABO system in this respect.
   d **False.** Normally Rh-negative people have no anti-D present.
   e **False.** The Kleihauer test is used to detect cells containing fetal Hb when screening maternal blood films.

34 a **False.** If the father is heterozygous Dd, the fetus may be dd and thus avoids rhesus disease.
   b **False.** A first baby may escape rhesus disease if the mother has not been sensitized previously.
   c **True.** The number of fetal cells in maternal blood rises especially between 18 and 22 weeks of pregnancy.
   d **False.** A mother may already be sensitized, e.g. by a Rh-positive transfusion.
   e **False.** A feto-maternal transfusion of up to 15 ml has been found to occur in about 0.3% of women.

Blood clotting and associated disorders (Qs 35–49)

35 *Tissue factor (TF):*
   a TF is expressed on the membrane of activated platelets.
   b TF is the key to activation of the extrinsic clotting system.
   c VIIa/TF stimulates the intrinsic system by activating XI to XIa.
   d VIIa/TF stimulates the common pathway by activating X to Xa.
   e VIIa/TF is self-generating.

36 *In blood coagulation:*
   a Factor XII is activated during the contact phase of blood clotting.
   b Factor XII is essential for normal clotting *in vivo*.
   c Factor XI deficiency is typically associated with a bleeding diathesis.
   d Factors VIII and V are proenzymes.
   e All coagulation reactions occur on cell surfaces.

37 *Thrombin:*

a   is a weak proteolytic enzyme.
b   activates many of the proenzymes in the clotting cascade.
c   is strongly inhibited by heparin in normal physiology.
d   is both a thrombotic and an antithrombotic agent.
e   normally helps to limit the extent of a clot.

38 *Fibrinogen:*

a   is a very large plasma protein.
b   levels in plasma fall when taking oral contraceptives (OCs).
c   is an independent risk factor for CVS disease if levels are persistently raised.
d   and its metabolites can damage the vascular endothelium.
e   concentration is raised after a stroke and after a myocardial infarct.

39 *Vitamin K:*

a   must be oxidized in order to be active physiologically.
b   formation (of the active form) involves carboxylation of glutamic acid.
c   dependent proteins must contain γ-carboxyglutamate residues (Gla residues) in order to be biologically active.
d   deficiency is almost never suffered by bottle-fed babies.
e   deficiency's most dangerous aspect in the neonate is intracranial bleeding.

40 *Vitamin K-dependent proteins:*

a   Factors II, V, VII, IX, and X are all vitamin K dependent.
b   only occur in the clotting cascade.
c   Tissue factor is vitamin K dependent.
d   Regulatory proteins S and C are vitamin K dependent.
e   Factor II is the most important vitamin K-dependent protein in blood clotting.

41 *Platelets:*

a   differentiate by endoreduplication rather than by mitosis.
b   production is controlled by a single haematopoietic growth factor.
c   adhere to subendothelial von Williebrand factor (vWF).
d   use considerable metabolic energy during adhesion.
e   adhesion provokes aggregation, swelling, and degranulation.

42 *With platelets:*

a   Aggregration receptors (AgRs) are different to adhesion receptors.
b   AgRs of platelets bind to one specific ligand.
c   The manufacture of considerable amounts of serotonin (5-HT) occurs.
d   Vesicles release 5-HT by a calcium-dependent mechanism.
e   The phospholipid is a rich source of prostacyclin (PGI_2).

43 *In tests of haemostatic function:*

a   The addition of ADP and adrenaline to platelet-enriched plasma causes platelet aggregation in normal people.

b   The normal bleeding time is between 2 and 9 minutes.

c   The prothrombin time (PT) can be used to evaluate the integrity of the intrinsic coagulation system.

d   The activated partial thromboplastin time (APTT) evaluates the integrity of the intrinsic system plus the common pathway.

e   The clotting factors must fall to near 50% of normal before the coagulation (clotting) time becomes prolonged.

44 *Tests in haematology:*

a   The erythrocyte sedimentation rate (ESR) increases in hyperfibrinogenaemia.

b   The ESR is increased by the presence of HbS.

c   The haematocrit level (cells/plasma) increases in burnt patients.

d   The Schilling test involves giving oral and IM radiocobalt approximately simultaneously.

e   Total iron binding capacity (TIBC) and serum Fe are low in Fe deficiency.

45 *Anticoagulation:*

a   Antithrombin III is a powerful *in vivo* anticoagulant.

b   Endothelial prostacyclin (PGI_2)–platelet thromboxane A_2 (Tx A_2) balance is important in maintaining the fluidity of blood.

c   Thrombin, streptokinase, and urokinase are all plasminogen activators (PAs).

d   *In vivo*, PAs are inhibited by physiological inhibitors (PAIs).

e   Activated protein C (APC) inactivates coagulation cofactors Va and VIIIa.

46 *Heparin:*

a   therapy can cause a thrombocytopenia and thrombosis as a side-effect.

b   exerts its coagulant effect by working as a cofactor for antithrombin III.

c   is absorbed well by the normal small intestine.

d   is a heterogeneous mucopolysaccharide.

e   readily crosses the placental barrier.

47 *In coagulation/anticoagulation balance:*

a   Protein S is a cofactor for plasmin.

b   Protein S is vitamin K dependent.

c   Hereditary resistance to protein C is due to factor V Leiden mutation.

d   Thrombophilia is a consequence of a deficiency of protein C, protein S, or antithrombin III.

e   The vast majority of homozygous autosomal coagulopathies appear clinically before 10 years of age.

48 *Anticoagulation therapy:*

    a Vitamin K antagonists are slowly acting *in vitro* anticoagulants.

    b Warfarin directly inhibits prothrombin synthesis in the liver.

    c Aspirin blocks prostacyclin formation by the endothelial cell.

    d Aspirin irreversibly blocks TxA_2 synthesis in the platelet.

    e Dextrans are glucose polymers that inhibit platelet function.

49 *Blood dyscrasias:*

    a Purpura describes any condition with an abnormal tendency to bleeding.

    b Haemophilia A is the most common hereditary coagulation disorder.

    c Haemophilia A is due to a defective functioning of factor IX.

    d Haemophilia B is due to a defective factor X.

    e von Willebrand's disease is an autosomal dominant disease.

Blood clotting and associated disorders (Answers)

35   a **False.** TF is found predominantly on the membrane of extravascular cells and also on the membrane of activated monocytes.

    b **True.** The extracellular domain of TF is a factor VII receptor.

    c **False.** VII/TF activates factor IX to factor IXa.

    d **True.** VIIa/TF rapidly promotes the generation of thrombin.

    e **True.** Factor Xa and thrombin both augment VIIa/TF formation, which in turn promotes further Xa generation.

36   a **True.** Factor XIIa is the classical initiator of the intrinsic system.

    b **False.** A deficiency of factor XII has little or no effects *in vivo*.

    c **False.** Neither factor XI nor factor XII are crucial for *in vivo* thrombogenesis.

    d **False.** Factors VIII and V are procofactors unrelated to the classic clotting proenzymes.

    e **True.** Clotting reactions occur on cell surfaces especially in the presence of plasma cofactors (calcium, VIIIa, Va).

37   a **False.** Thrombin is a large and powerful protease. Classically it cleaves soluble fibrin from fibrinogen.

    b **True.** Thrombin acts at all stages of the clotting cascade.

    c **False.** On its own heparin is a very weak antithrombin.

    d **True.** Antithrombotic action of thrombin is apparent when it is attached to thrombomodulin of the endothelial wall.

    e **True.** Thrombin limits clot extension by its antithrombotic action at either end of the clot.

38   a **True.** Fibrinogen is one of the largest plasma proteins. It is present in a concentration of 0.2–0.3 g/dl.

    b **False.** OCs, especially oestrogen-containing OCs, raise fibrinogen levels.

c **True.** Fibrinogen and von Willebrand factor are the principal adhesive macromolecules that link platelets together. They bind with the aggregatory platelet receptor IIb/IIIa (one of the integrin families of protein).

d **True.** Fibrinogen and von Willebrand factor are prethrombotic and may damage the endothelium by releasing platelet products. Homocysteine also damages the vessel wall.

e **True.** As an acute-phase protein fibrinogen is elevated in inflammation and necrosis. However, it is also often raised before these events.

39 a **False.** Only the reduced form (KH_2) is active. KH_2 is made in the liver and needs NADH for its formation.

b **False.** The oxidation of vitamin KH_2 (active) to vitamin KO (inactive) creates γ-carboxyglutamates (Gla residues) in vitamin K-dependent proteins.

c **True.** Gla is necessary for reactions dependent on calcium ion binding.

d **True.** 25–50 mg vitamin K taken daily in formula or cow's milk is sufficient to prevent vitamin K deficiency.

e **True.** Vitamin K deficiency causes generalized bleeding but specifically intracranial bleeding is dangerous in the neonate.

40 a **False.** Factor V does not contain the necessary Gla and propeptide domain.

b **False.** Vitamin K-dependent proteins are also found in mineralized tissues, e.g. bone, dentin.

c **False.** TF contains neither Gla nor a propeptide domain.

d **True.** Proteins S and C are important in controlling the rate and extent of thrombosis.

e **True.** In general, this is true. Note that prothrombin contains 10 tightly clustered Gla residues.

41 a **False.** Megakaryocytes endoreduplicate forming polyploid cells. Platelets do not divide.

b **False.** Cofactors promoting platelet production include IL-3, IL-6, IL-11, and granulocyte–macrophage colony-stimulating factor (GM-CSF).

c **True.** Platelets adhere to endothelial substances, e.g. vWF, collagen, fibronectin, and thrombospondin, via platelet 1b membrane adhesion receptors.

d **False.** No platelet metabolic activity is required for platelet adhesion to occur.

e **True.** Aggregation of platelets allows expression of receptors for factor V, cofactor VIII, and for a P-selectin receptor for leucocytes. Platelet metabolic energy is needed for this to occur.

42 a **True.** Adhesive receptors can be either integrin or non-integrin in nature. Both types are platelet membrane glycoproteins. The integrin glycoprotein IIa/IIIb receptor is specific to the platelet and megakaryocyte.

b **False.** Many ligands exist for platelet aggregation receptors, e.g. fibrinogen, vWF, fibronectin, thrombospondin.

c **False.** The platelet picks up 5-HT (serotonin) from the plasma and stores it.

d **True.** 5-HT is stored in vesicles similar to those in nerve endings.

e **False.** Thromboxane A_2 (TxA_2) is the chief prostaglandin of platelets.

43 a **True.** Reversible platelet aggregation occurs. This later becomes irreversible due to the release of TxA_2 and other platelet products.

b **True.** The bleeding time is a crude guide to platelet function. It is prolonged in anaemia.

c **False.** PT is the time required for fibrin strands to appear after adding TF and Ca^{2+} to platelet-poor plasma. Normally this takes 11–13 s. PT can be used to evaluate both extrinsic and common pathways.

d **True.** The APTT: kaolin/cephalin/PO_4 plus Ca^{2+} is added to plasma. Fibrin should normally appear in 23–31 s.

e **False.** Coagulation may be virtually normal even if the factors are only 5% of normal. Usual coagulation time is 6–12 min.

44 a **True.** The effect on ESR, out of scale of 10: for fibrinogen, 10; beta-globulin, 5; gammaglobulin, 2; alphaglobulin, 2; albumin, 1.

b **False.** The abnormal shape of HbS interferes with rouleaux and sedimentation rate .

c **True.** Haematocrit increases because of the loss of plasma at the burnt area.

d **True.** Urinary excretion of radiocobalt is < 5% in pernicious anaemia.

e **False.** In iron deficiency, TIBC > 350 μg/dl; serum iron < 35 μg/dl.

45 a **False.** Antithrombin III needs heparin as a cofactor. Factor Xa stimulates antithrombin III.

b **True.** PGI_2/TxA_2 balance may be altered in disease, e.g. Prinzmetal's angina.

c **True.** They all act as primary plasminogen activators mainly by the release of secondary activators from the vascular endothelium.

d **True.** Haemorrhage is caused by lack of PAI-1 as in a congenital lack or in the haemolytic uraemic syndrome.

e **True.** Thrombin/thrombomodulin complex generates activated protein C (APC).

46 a **True.** Hypothesis: heparin–platelet factor 4 (PF4)–Ig complexes damage platelets. This promotes bleeding. Excess PF4 damages endothelium, thus encouraging thrombosis.

b **False.** Heparin/antithrombin III inhibit thrombin activation of factor V and factor VIII, that is why heparin has an anticoagulant effect.

c **False.** Heparin must be given IV or SC. Half-life is 60–90 min.

d **True.** Heparin varies in mol. wt from 2000 to 40 000 ('small' to 'large' heparin).

e **False.** The failure to cross the placenta makes heparin the anticoagulant of choice in pregnancy.

47   a   **False.** Protein S is a cofactor for protein C.

    b   **True.** Protein S and protein C are both vitamin K dependent.

    c   **True.** Factor V Leiden mutation occurs in about 50% of patients with familial thrombophilia. This renders factor V resistant to inactivation by protein C.

    d   **True.** Homozygous state causes severe syndromes from infancy on.

    e   **True.** However, the timing of the clinical onset of hereditary coagulopathies is affected by sex, incomplete penetrance of the gene, pregnancy, sepsis, trauma, and drugs.

48   a   **False.** Vitamin K antagonists only act *in vivo*.

    b   **False.** Warfarin blocks reactivation of inactive vitamin KO to active KH_2.

    c   **True.** Endothelial cells regenerate cyclo-oxygenase and thus make new PGI_2.

    d   **True.** Platelets cannot regenerate cyclo-oxygenase.

    e   **True.** Dextrans inhibit platelet function. Dextran 40 kDa, half-life 3 h; dextran 70 kDa, half-life 24–28 h.

49   a   **True.** In purpura there are defects in one or more of: platelets, vessel wall, coagulation pathways.

    b   **True.** In one-third of cases of haemophilia A there is a recent genetic mutation.

    c   **False.** Haemophilia A is due to lack of factor VIII. Overt cases only occur in males (X-linked recessive gene).

    d   **False.** Haemophilia B (Christmas disease) is due to lack of factor IX.

    e   **True.** Von Willebrand's disease occurs in both sexes (contrast with haemophilia A).

Kinins and eicosanoids (Qs 50–58)

50 The kinins:

    a   are biologically active lipids derived from membrane phospholipid.

    b   are rapidly destroyed in the plasma by kininases.

    c   are generally strong vasoconstrictor agents.

    d   are produced by the effect of kallikrein on prekallikrein.

    e   act as plasminogen activators.

51 Eicosanoids:

    a   are small peptide mediators derived from arachidonic acid.

    b   are stored in the cells of origin.

    c   act chiefly on the cell of origin.

    d   diffuse into target cells.

    e   are found in abundance in lymphocytes.

52 Arachidonic acid (AA):

a   is a polyunsaturated fatty acid.
b   is an essential constituent of the diet.
c   is a constituent of the phospholipid component of cell membranes.
d   is only released by specific stimuli to the cell membrane.
e   is split from membrane phospholipids by lipoprotein lipase.

53 Metabolism of arachidonic acid (AA):

a   The type of eicosanoid released depends on the cell involved.
b   Cyclo-oxygenase is involved in the breakdown of arachidonic acid.
c   Lipoxygenase-mediated pathway leads to leukotriene formation.
d   The epoxygenase pathway leads to omega fatty acid formation (lipoxins).
e   Platelet activating factor (PAF) comes from the cyclo-oxygenase pathway.

54 Smooth muscle effects of prostaglandins (PGs):

a   Prostaglandins predominantly relax the smooth muscle of the gut.
b   Both PGI_2 and TxA_2 relax bronchial smooth muscle.
c   Prostaglandins predominantly relax the pregnant uterus.
d   Prostaglandins tend to close the ductus arteriosus (DA).
e PGD_2 is a powerful vasoconstrictor.

55 Lipoxygenase-derived metabolites (leukotrienes):

a   are made predominantly by leucocytes.
b   generally relax smooth muscle.
c   increase vascular permeability.
d   depress myocardial activity.
e   inhibit leucocyte chemotaxis.

56 Tissue-specific prostaglandins:

a   Renal prostaglandins are chiefly made in the renal cortex.
b   Renal PGs increase renin production.
c   In quiet, resting conditions renal PGs are very important in controlling renal blood flow.
d   Gastric PGs are ulcerogenic.
e   Brain PGs are involved in the development of fever.

57 Inhibition of eicosanoid synthesis:

a   Corticosteroids inhibit cyclo-oxygenase.
b   Non-steroidal anti-inflammatory drugs (NSAIDs) inhibit cyclo-oxygenase irreversibly.
c   Paracetamol (acetaminophen) is a potent inhibitor of cyclo-oxygenase.
d   The cyclo-oxygenase inhibitor sulphinpyrazone may spare renal PG synthesis.
e   NSAIDs increase the production of leukotrienes.

58 *The prostaglandin, prostacyclin, PGI$_2$:*

a is found chiefly in vascular endothelial cells.
b is both a vasodilator and antiplatelet agent.
c can be formed from TxA$_2$ by the endothelial cell.
d Pulsatile pressure stimulates its production.
e is produced more readily in those with diabetes mellitus.

Kinins and Eicosanoids (Answers)

50 a **False.** The kinins are highly active polypeptides containing 9–11 amino acids.
   b **True.** Note that ACE (angiotensin-converting enzyme) is kininase 11.
   c **False.** Kinins, e.g. bradykinin, are vasodilator agents.
   d **False.** Kallikrein converts kininogen to kinin and prorenin to renin. Tissue damage, factor XIIa, complement, amines, and some PGs convert prekallekrein to the enzyme kallikrein.
   e **True.** Thus the kinins are useful in the treatment of tissue damage and inflammation.

51 a **False.** Eicosanoids are lipid mediators. Their biosynthesis is limited by the availability of arachidonic acid.
   b **False.** Eicosanoids are used rapidly and are not stored.
   c **False.** Eicosanoids are formed in cells and released into local ECF.
   d **False.** Eicosanoids bind to specific receptors on target cell membrane.
   e **False.** Lymphocytes are unique; they produce little or no eicosanoid.

52 a **True.** Arachadonic acid (AA) is formed *in vivo* from linoleic and linolenic acid.
   b **False.** AA is essential only if the diet lacks linoleic and linolenic acid.
   c **True.** Any free arachidonate in cells is incorporated into cellular and membrane lipids.
   d **False.** Non-specific stimuli, e.g. hypoxia, pH change, may also cause the release of arachadonic acid.
   e **False.** Phospholipase does this. Endothelial lipoprotein lipase splits triglycerides into free fatty acid and glycerol.

53 a **True.** TxA$_2$ comes mainly from platelets, PGI$_2$ from the vascular endothelium.
   b **False.** Cyclo-oxygenase is involved in prostaglandin production from arachidonic acid.
   c **True.** Lipoxygenation reactions occur typically in neutrophils.
   d **True.** The epoxygenase pathway involves omega oxidation by cytochrome P450 enzymes of microsomes.
   e **False.** PAF comes from reacylation of free AA. PAF is a potent pathophysiological mediator in asthma and shock.

54 a **False.** PGF_2, PGI_2, PGE_2 contract GIT via the release of calcium.

   b **False.** TxA_2 and PGF vasoconstrict and bronchoconstrict. PGI_2 and PGE_2 vasodilate and bronchodilate (via cAMP).

   c **False.** PGF and PGE are powerful stimulants of the pregnant uterus.

   d **False.** PGE_2 and PGI_2 keep the ductus arteriosus open. PG inhibitors, e.g. NSAIDs, close it.

   e **False.** PGD_2 is found mainly in mast cells. It has a possible role in allergic asthma, immunology, and in normal sleep.

55 a **True.** Leukotrienes are so-called because they are made in abundance by leucocytes.

   b **False.** Leukotrienes generally produce smooth muscle spasm.

   c **True.** Thus leukotrienes are useful in the inflammatory process.

   d **True.** Increased levels of leukotrienes occur in shock, trauma, cardiac ischaemia, respiratory distress syndrome, inflammatory disease.

   e **False.** Leukotrienes attract leucocytes to the site of release.

56 a **False.** Renal PGs are made chiefly in the renal medulla.

   b **True.** Renal PGs increase renin and PGI_2. None the less, sympathoadrenal activity is a more important stimulus.

   c **False.** Renal PGs assume importance in controlling renal circulation in conditions of vasoconstriction.

   d **False.** Gastric PGs appear to be cytoprotective for gastric mucosa.

   e **True.** The PGs that mediate fever are not yet known.

57 a **False.** Steroids induce a phospholipase-inhibitory protein (lipocortin).

   b **False.** Only aspirin (which is not classified as an NSAID) inhibits cyclo-oxygenase irreversibly.

   c **False.** Paracetamol is a weak inhibitor of cyclo-oxygenase, except perhaps, in the brain.

   d **True.** Because renal tissue reconverts pyrazones to inactive sulphones.

   e **True.** NSAIDs divert arachidonic acid from the cyclo-oxygenase pathway to the lipoxygenase pathway.

58 a **True.** PGI_2 is also made in small amounts elsewhere, e.g. in the platelets.

   b **True.** Unfortunately PGI_2 has not fulfilled therapeutic hopes so far.

   c **True.** Some PGI_2 is normally made in the endothelial cell from TxA_2 that diffuses from freely circulating platelets.

   d **True.** PGI_2 production is also stimulated by kinins, thrombin, 5-HT, TxA_2, and ADP.

   e **False.** PGI_2 synthesis decreases in diabetes mellitus, with age, and in cigarette smokers.

3

Immunology

Complement (Qs 1–2)

1 Complement proteins (C-proteins):

a are produced in significant amounts in the epithelium of the small intestine.

b are produced in significant amounts by macrophages and fibroblasts.

c include C-3 which is chiefly produced in the liver.

d are acute-phase proteins.

e in their synthesis both classical and alternative routes provide C3 convertase.

2 Regarding the complement system:

a The alternative pathway is activated by both immune and non-immune factors.

b Properdin is essential for activation of the alternative pathway.

c C5b–C9 complexes deposit on the surfaces of cells and bacteria.

d Paroxysmal nocturnal haemoglobinuria (PNH) is due to lack of a C-protein.

e 'First time' hypersensitive reactions may be due to C-proteins.

Complement (Answers)

1 a **True**. The GIT epithelium is a rich source of complement, especially of the C1 class.

b **True**. Macrophages and fibroblasts synthesize C2, C3, C4, C5, and factor B.

c **True**. The synthesis of C3 occurs predominantly in the liver.

d **True**. C1 inhibitor, factor B, C2, and C3 rise sharply in inflammatory states. C reactive protein is the major acute phase protein.

e **True**. C3 convertase converts C3 to C3b.

2 a **True**. Igs as well as properdin, endotoxin, and complex saccharides activate the alternative pathway.

b **False**. C3, B, and D are early components of the alternative pathway. Properdin augments only.

c **True**. The deposition of complement complexes on the surface of cells facilitates cytolysis.

d **False**. In PNH the red cells bind more C3b than normal and are hypersensitive to the cytolytic effects of complement, especially in acidic or hypoxic conditions.

e **True**. First-time hypersensitivity often shows massive activation of the alternative system (see reactions to non-biocompatible dialysis membranes).

Granulocytes, macrophages, and lymphocytes (Qs 3–9)

3 Neutrophil granulocytes:

a Most of the circulating leucocytes are neutrophil granulocytes.

b The normal ratio of marrow to intravascular neutrophils is about 50:1.

c Exercise is associated with a leucocytosis.

d Band forms of neutrophils (unsegmented nuclei) are found in the peripheral blood in exercise.

e Fully differentiated granulocytes are microbiocidal and phagocytic.

4 Concerning monocytes and macrophages:

a Each monocyte only circulates for about 24–72 h in the blood.

b Some monocytes settle extravascularly and become tissue macrophages.

c Macrophages play a central role in controlling haematopoiesis.

d Küpffer cells (lining liver sinusoids) are derived from blood monocytes.

e Blood monocytes give rise to mast cells.

5 Eosinophils:

a are found in equal numbers intra- and extravascularly.

b when mature, have a multilobed nucleus.

c acquire increased nuclear lobulation with age.

d increase in the blood in acute bacterial infections.

e contain large amounts of histamine.

6 Concerning eosinophil function:

a Major basic protein (MBP) is found abundantly in the nucleus of the eosinophil.

b Degranulation of the eosinophil helps augment mast cell effects.

c MBP is toxic to helminths and to some bacteria.

d MBP can damage the human host.

e Eosinophils carry receptors for immunoglobulins.

7 Basophils:

a have a bilobed nucleus.

b contain cytoplasm packed with small purple metachromatic granules.

c are the precursors of tissue mast cells.

d have a histological appearance very similar to that of mast cells.

e carry surface membrane receptors for the Fc fragment of IgE.

8 Mast cells:

a allow a local build up of IgE.

b degranulate in response to a local build up of IgE.

c along with basophils contain mediators typical of histamine anaphylaxis.

d   contain the same mediator content whatever their location.

e   are very often involved in late-phase reactions at sites of allergic challenge.

9 *About lymphocytes:*

a   Marrow stem cells give rise to precursors of T and B cells.

b   B cells are more numerous in peripheral blood than T cells.

c   B cells form typical whorls ('rosettes') with normal sheep RBCs.

d   Lymphocytes only spend a few hours in the bloodstream at any one time.

e   Plasma cells are derived from activated B cells.

Granulocytes, macrophages, and lymphocytes (Answers)

3   a   **True.** Usual differential count (% leucocytes): neutrophils 60–70%, eosinophils 2–4%, basophils 0.5–1%, monocytes 3–8%, lymphocytes 20–25%.

   b   **False.** The marrow:intravascular neutrophil ratio is about 3:1.

   c   **True.** Increased numbers of granulocytes are released from the marrow pool in exercise.

   d   **False.** Immature WBCs (band forms) suggest a pathological cause.

   e   **True.** Unlike monocytes which must undergo further differentiation before they are microbiocidal and phagocytic.

4   a   **True.** The majority of monocytes are destroyed in the bone marrow after circulating for 24–72 h in the blood.

   b   **True.** Some monocytes settle in extravascular sites, particularly in the liver, spleen, bone marrow, lungs, and lymph nodes.

   c   **True.** Macrophages secrete IL-1, TNF, and colony-stimulating factors (CSFs), all of which are powerful haematopoietins.

   d   **True.** Küpffer cells are part of the tissue mononuclear phagocyte system.

   e   **False.** Mast cells have a distinct origin in the marrow. IL-3 stimulates their formation.

5   a   **False.** Eosinophils are about 100 times more numerous in extravascular tissues, especially in the skin and in the submucosa of the respiratory system (RS), GIT, and urinary tract.

   b   **False.** A mature eosinophil has a bilobed nucleus.

   c   **False.** For all granulocytes the lobulation depends on the progenitor cell and is fixed for the cell's life span. Note that there is increased lobulation of neutrophils in pernicious anaemia.

   d   **False.** Eosinopenia is the rule in infections. Cortisol also causes eosinopenia.

   e   **False.** Eosinophils contain even more histaminase than neutrophils.

6 a **False.** Arginine-rich MBP is confined to the red-staining granules of the eosinophil.
   b **False.** Eosinophil products inactivate mediators from mast cells.
   c **True.** Eosinophilia is associated with helminthiasis.
   d **True.** MBP has a pathogenic role in hypereosinophilic syndromes.
   e **True.** Eosinophils also carry receptors for complement, cytokines, and eicosanoids.

7 a **False.** Basophils have a multilobed nucleus.
   b **False.** Basophil cytoplasm is packed with large purple metachromatic granules.
   c **False.** Mast cells come from a special bone marrow progenitor cell.
   d **False.** Mast cells have a distinctive round nucleus.
   e **True.** Both basophils and mast cells have membrane receptors for IgE.

8 a **True.** The external surface of the mast cell has up to 500 000 receptors for IgE.
   b **True.** When IgE reaches a certain level degranulation of mast cells occurs.
   c **True.** Activated mast cells and basophils also release heparin, eicosanoids (PGs, PAF), cytokines, and proteases.
   d **False.** There are substantial differences between mast cells in content, sensitivity, and appearance in different microenvironments of the body.
   e **True.** Late-phase reactions are probably due to enhanced vascular permeability and leucocyte infiltration at sites of allergen challenge.

9 a **True.** Precursors of T and B cells are processed into immunocompetent cells in the thymus and marrow, respectively.
   b **False.** T cells constitute 70% of circulating lymphocytes.
   c **False.** T cells form 'rosettes' with sheep red blood cells.
   d **True.** Both B and T cells recirculate between peripheral lymphoid tissue and blood many times a day.
   e **True.** Plasma cells are capable of intense protein synthesis.

Cytokines (Qs 10–12)

10 Cytokines:
   a are substances which are distinct from lymphokines.
   b are soluble bioregulatory polypeptides.
   c are predominantly local hormones (autocrine) in tissue of origin.
   d may act as poietins (growth-promoting factors).
   e link immune and endocrine systems in adrenal cortex.

11 *The following are effects of cytokines:*

a stimulation of the pituitary gonadal axis.
b an overall increase in body glucose turnover.
c stimulation of proteolysis, e.g. in skeletal muscle.
d induction of fever, anorexia, cachexia.
e a fall in the serum iron concentrations

12 *Interleukin 1 (IL-1):*

a family contains three related polypeptides.
b is secreted mainly by lymphocytes.
c stimulates the release of many pituitary hormones.
d and other cytokines are secreted by microglia of the brain.
e stimulates T and B lymphocytes.

Cytokines (Answers)

10 a **False.** Originally thought that lymphocytes alone produced cytokines, hence the old name 'lymphokine'.

b **True.** Cytokines are secreted mostly by activated T lymphocytes and macrophages.

c **False.** The effects of cytokines are often widespread and general.

d **True.** Many cytokines act as poietins, e.g. GM-CSF (granulocyte–macrophage colony-stimulating factor).

e **False.** Cytokines probably link the immune and endocrine systems at the hypothalamo-pituitary axis.

11 a **False.** Cytokines, e.g. TNF, inhibit the pituitary–gonadal axis.

b **True.** Cytokines increase the production and tissue uptake of glucose (independently of insulin).

c **True.** Cytokines also stimulate acute phase protein synthesis in the liver, e.g. IL-1, IL-6, and TNF.

d **True.** The old name for tumour necrosis factor was cachectin.

e **True.** Fe and Zn fall, Cu may rise. Possibly these changes help inhibit bacterial growth.

12 a **True.** The interleukin-1 family consists of: IL-1a; IL-1b and the interleukin-1 receptor antagonist (IL-1 RA).

b **False.** IL-1 is secreted mainly by monocytes that are already stimulated by microbial products or inflammation.

c **True.** IL-1 is especially powerful in stimulating corticotrophin and vasopressin.

d **True.** All immunocompetent cells can secrete cytokines.

e **True.** The positive effect of IL-1 on lymphocytes has been shown to reduce mortality from bacterial infections in animals.

Immunoglobulins (Qs 13–16)

13 *Regarding the structure of immunoglobulins (Igs):*
a The basic unit is made up of four identical polypeptide chains.
b Igs can be split into Fab and Fc fragments by papain digestion.
c The Fab fragment has a variable (V) structure.
d The Fab fragment binds complement and macrophages.
e There is a hypervariable region within the variable region.

14 *Immunoglobulins (Igs):*
a Igs are divided on the basis of differences in their light chains.
b IgG is the lightest and most abundant immunoglobulin type.
c IgG is the first Ig secreted in most cases of antigen re-exposure.
d IgG is the only immunoglobulin that can cross the human placenta.
e A resting B-cell has an antibody of one specificity on its surface.

15 *In immunology:*
a An idiotype is an external antigen after it has been altered by the immune response.
b An accessory cell is any cell that can capture antigen.
c An epitope is a small piece of detached antibody.
d CD8 is a suppressor cytotoxic T-cell.
e Genes in the major histocompatibility complex (MCH) control immuno-competent cells.

16 *In immune function:*
a B cells cannot bind directly to unprocessed antigen.
b T cells can only bind to antigen processed by macrophages.
c activated T-cells produce Igs in abundance.
d B cells are chiefly activated by cytokines from T-cells.
e the helper CD4:suppressor CD8 ratio is less than unity in HIV infection.

Immunoglobulins (Answers)

13 a **False.** The basic unit of Ig consists of two identical heavy (H) polypeptide chains and two identical light (L) polypeptide chains.

b **True.** Papain digestion splits the Fc 'handle' (all H chain) from the rest of the molecule.

c **True.** The Fab fragment always contains the NH_3 terminals of the H and L chains.

d **False.** Fab binds antigen, Fc binds complement and macrophages.

e **True.** The folds of each V domain form three short loops of amino acid.

14 a **False.** Igs are divided on differences in H chains into IgG, A, M, E, D.
   b **True.** IgG is about 150 000 daltons; the heaviest is IgM, about 900 000 daltons.
   c **True.** IgM is secreted on first antigenic exposure. This shifts to IgG or IgA on re-exposure.
   d **True.** IgG confers passive immunity for many diseases. Note that anti-D is an IgG.
   e **True.** At any one time about 10^8 different clones of B cells are present in the body, each clone displaying a unique antibody.

15 a **False.** Idiotypes are the antigenic determinants of the variable region of the Fab fragment. In high concentration an idiotype can induce anti-idiotypes (autoimmunity).
   b **True.** Activated monocytes–macrophages, neutrophils, and histiocytes are all examples of accessory cells.
   c **False.** An epitope is a small antigenic fragment presented by an accessory cell to a passing lymphocyte.
   d **True.** A CD4 cell is a helper T-cell.
   e **True.** Class I, II, and III genes of MCH code appropriate antigen trapping and recognition molecules on T-, B-, and accessory cells.

16 a **False.** A few antigens can short circuit macrophage processing and bind directly to B cells.
   b **True.** Antigen is processed by macrophage and then presented to resting T cells along with IL-1 and a class II MCH molecule.
   c **False.** Activated T-cells produce cytokines in abundance.
   d **True.** A battery of T-cell cytokines activate the B cells.
   e **True.** Normally the CD4:CD8 ratio is about 2.0:1.0. It is usually < 1.0 in HIV infection.

Anaphylaxis (Q17)

17 *Anaphylaxis:*
   a This is a term used to describe allergic-type reactions other than those mediated by IgE.
   b Histamine is an important mediator of acute anaphylaxis.
   c Antihistamines are effective in treating anaphylaxis.
   d Adrenaline is contraindicated in treating children with anaphylaxis.
   e Primary cardiac failure is a very frequent cause of hypotension in severe anaphylaxis.

Anaphylaxis (Answers)

17 a **False.** Anaphylaxis is mediated by IgE or IgG. Anaphylactoid reactions are
not mediated by IgE.

b **True.** Histamine is an early mediator, but normal plasma concentrations
are often found within minutes of the reaction occurring.

c **True.** Adrenaline, antihistamines, steroids, and volume replacement are all
used. Adrenaline is the initial preferred treatment by most clinicians, but
most would use all if required.

d **False.** For children adrenaline is given in a dose of 0.01 ml/kg of a 1:1000
solution IM or 0.01 ml/kg of a 1:10 000 solution IV.

e **False.** The hypotension of anaphylaxis is mostly due to vasodilation and
plasma loss and not to heart failure.

4

Renal

Renal structure and pressures (Qs 1–3)

1 *In the kidney:*

a there are approximately as many cortical as juxtamedullary glomeruli.
b the blood flow is the highest per gram of any organ in the body.
c the arteriovenous oxygen difference is greater than in most other organs.
d the oxygen usage rate (ml O_2/min) is second only to that of the heart.
e medullary blood flow represents only 1–2% of total renal blood flow (RBF).

2 *In the renal vascular system:*

a Glomerular capillary pressure is approximately the same as the arterial diastolic BP.
b Hydrostatic pressure is higher in glomerular capillaries than in non-renal ones.
c Glomerular filtration of a substance does not occur if its mol. wt > than 70 000 (approx.).
d Autoregulation of renal blood flow is maintained over a BP range of 50–200 mmHg.
e Autoregulation is disturbed in sympathectomized kidneys.

3 *Glomerular filtration:*

a ceases at a hydrostatic pressure in Bowman's capsule of approximately 10 mmHg.
b ceases when the oncotic pressure in the glomerular capillaries reaches about 5 mmHg.
c normally accounts for about about one-fifth of the blood volume that perfuses the glomerular capillaries.
d is improved by IgA deposits in the glomerulus.
e indexes the final urine volume in a fairly linear way.

Renal structure and pressures (Answers)

1 a **False.** In man only about 15% of glomeruli are juxtamedullary.
   b **False.** Carotid bodies have highest blood flow per gram. Note the total RBF is approximately 1300 ml/min.
   c **False.** The kidneys extract only about 1.5 ml of O_2 per 100 ml renal blood flow.
   d **True.** Renal oxygen usage is about 15 ml/min, second only to the heart (25 ml/min).
   e **True.** Medullary flow is small (10–21 ml/min).

2 a **False.** Glomerular capillary pressure is 35–45 mmHg or about 40% of the mean aortic pressure.

b **True.** High glomerular capillary pressure is due to the unique interposition of these vessels in the course of the renal circulation.

c **False.** Uncharged molecules up to 100 000 mol. wt can be filtered. Diffusion ceases around 70 000 for charged particles, e.g. albumin (a negatively charged elongated molecule).

d **False.** Autoregulation is compromised outside a BP range of 80–180 mmHg in normal people.

e **False.** Autoregulation is independent of nerves. It is essentially an intrinsic myogenic response.

3 a **False.** The pressure in Bowman's space is normally about 10 mmHg.

b **True.** However, the oncotic pressure in the glomerular capillaries never rises to this level in normal people because of the relatively small amount of filtrate formed in humans.

c **False.** Normal filtration fraction is 0.2 of the plasma that perfuses the glomerular capillaries.

d **False.** IgA nephropathy is the most common form of glomerulonephritis.

e **False.** Urine volume is controlled by the extent of the tubular reabsorption of water from the filtrate

Proteinuria and clearance (Qs 4–9)

4 Proteinuria:

a In health one normally loses 10-15 mg albumin/day in the urine.

b Normal adults can excrete up to 150 mg of protein/day in urine.

c Albuminuria of over 20 μg/min is always pathological.

d Orthostatic albuminuria may cause the loss of 2.0 g albumin/day.

e Bence Jones protein is the pure Ig of myelomatosis.

5 Concerning renal clearance (C):

a Renal clearance of substance X is expressed as the amount of X cleared from the blood in 1 min of blood flow through the kidneys.

b The glomerular filtration rate (GFR) is close to the C of inulin .

c Free water clearance is normally equal to the volume of urine.

d The clearance of bicarbonate is close to the GFR.

e Clearance of urea is unusual in that it rises as urine flow rate rises.

6 Renal clearance (C):

a of inulin increases as one gets older.

b of inulin is about 125 ml/min in a young adult (corresponds to 180 litres of filtrate/day).

c of glucose is 100%.

d of mannitol can be used to estimate the GFR.

e of para-aminohippurate (PAH) is roughly the same as the renal plasma flow.

7 *Creatinine:*

a   is derived almost entirely from muscle creatine.
b   clearance is slightly greater than that of inulin.
c   clearance is a more accurate assessment of GFR than the clearance of inulin in a clinical setting.
d   levels in the plasma are independent of dietary protein.
e   clearance may be falsely elevated in renal disease.

8 *The glomerular filtration rate (GFR):*

a   decreases in the erect position.
b   approaches zero if arterial blood pressure falls to less than 60 mmHg.
c   increases significantly in pregnancy.
d   increases in exercise.
e   increases on exposure to a cold environment.

9 *Concerning the afferent and efferent arterioles:*

a   'Myogenic response' to pressure change is abolished if myocytes are poisoned.
b   Myocytes of efferent arterioles are independent of circulating vasoactive metabolites.
c   The afferent arterioles are unusual in that their calibre is unaffected by prostaglandins.
d   The calibre of each type is synchronously adjusted to maintain the GFR.
e   The calibre is controlled by the solute load in the thick ascending limb of Henle's loop.

Proteinuria and clearance (Answers)

4   a   **True.** Small amounts of albumin manage to pass the glomerular sieve.
   b   **True.** Most protein in the urine comes from glycoproteins of the renal tubular cells.
   c   **True.** Microalbuminuria is typical of early diabetic nephropathy.
   d   **True.** Orthostatic albuminuria is possibly due to hypoxia of the glomerulus from renal vasoconstriction when a subject assumes the upright position.
   e   **False.** Bence Jones protein is the dimer of the L chain of the excess Ig of myelomatosis.

5   a   **False.** Clearance is expressed in ml of plasma cleared per unit time.
   b   **True.** Inulin is freely filtered and is neither secreted nor reabsorbed by the kidney tubules.
   c   **False.** Free water clearance is the volume of plasma from which solute-free water has been cleared in 1 min.
   d   **False.** Clearance of bicarbonate – 0.5 ml/min; that is, very little plasma is cleared of bicarbonate/min.

    e **True.** Clearance of urea varies from 54–74 ml/min, depending on the rate of urine formation

6 a **False.** The GFR normally falls by 1 ml/year after 35 years of age.

    b **True.** The GFR of each nephron is reckoned to be about 90 μl/day.

    c **False.** Clearance of glucose is 0 ml of plasma/min.

    d **True.** Mannitol, sorbitol, and hexitol are handled like inulin.

    e **True.** PAH and diodrast are freely filtered and are secreted by renal tubules and not reabsorbed. Clearance is around 600 ml/min.

7 a **True.** Creatinine is an organic base formed during muscle protein metabolism as a degradation product of muscle creatine phosphate.

    b **True.** This is because a small amount of creatinine is secreted in the proximal tubules in humans.

    c **False.** However, the C_{cr} is easier to do than the $C_{insulin}$ because there is no need to set up an infusion.

    d **False.** The production of creatinine in the body is largely, but not completely, independent of diet or exercise.

    e **True.** The tubular secretion of creatinine can rarely increase in renal failure.

8 a **True.** The GFR falls in the erect position because of the fall in pressure in the glomerular capillary consequent to vasoconstriction of the afferent arteriole which results from sympathoadrenal activation on standing.

    b **True.** The GFR ceases in severe hypotension and anuria results.

    c **True.** The GFR rises 20% in pregnancy. Probably associated with hydraemia.

    d **False.** Glomerular pressure falls because blood is diverted to the exercising muscles.

    e **False.** Glomerular pressure falls from sympathoadrenal renal vaso-constriction in a cold environment (see 8a).

9 a **True.** The myogenic relaxation/constriction response is intrinsic (is a property of the vascular smooth muscle).

    b **False.** Both types of renal arteriole are very sensitive to vasoactive agents, e.g. amines, angiotensin II.

    c **False.** Local endothelial PGs and renal medullary PGs are important in controlling the calibre of renal arterioles.

    d **True.** The relative diameter of afferent to efferent arterioles is a critical control of filtration pressure and hence of the GFR.

    e **True.** If the solute load falls, the macula densa responds by inducing relaxation of the afferent arteriole so that the GFR increases. This is tubuloglomerular feedback. PG mediators may be involved.

Tubule function (Qs 10–16)

10 At the proximal convoluted tubule (PCT):

a absorption occurs due to a 'leaky' epithelium.

b up to half of the filtered salt and water is reabsorbed in the PCT.

c luminal sodium ion is exchanged for cellular hydrogen ion in the PCT.

d the sodium pump on the luminal membrane helps in sodium reabsorption.

e reabsorption is affected by hydrostatic pressure in peritubular capillaries.

11 In the proximal convoluted tubule (PCT):

a The colloid osmotic pressure (COP) in the peritubular capillaries affects reabsorption in the PCT.

b Proximal tubular fluid reabsorption is strongly influenced by the GFR.

c Maximum glucose reabsorption is directly determined by the level of plasma glucose.

d Glucose is absorbed into luminal cells via a similar mechanism to that used by brain cells.

e Over 95% of inorganic phosphate is reabsorbed in the PCT.

12 Filtered solutes:

a Urea is reabsorbed both actively and passively in the nephron.

b All filtered potassium is normally reabsorbed in the PCT.

c Foreign materials, e.g. PAH and diodrast, are chiefly secreted by the distal convoluted tubule (DCT).

d Non-ionized calcium is reabsorbed in the PCT.

e Significant amounts of calcium are reabsorbed in parts of the nephron other than the PCT.

13 Other miscellaneous solutes:

a Urate is both secreted and reabsorbed in the nephron.

b Almost all the filtered bicarbonate is reabsorbed early in the PCT.

c Carbonic anhydrase (CA) is absent from the cells of the distal tubule.

d Most hydrogen ions are secreted in distal parts of the nephron.

e Most metabolic end products, e.g. adrenaline, glucuronides, and sulphates, are secreted by the PCT.

14 Henle's loop:

a Henle's loop is essential for the production of a concentrated urine (urine that is hypertonic to plasma).

b Isotonic fluid that enters the loop leaves in a hypotonic condition.

c Na^+ is actively secreted into the interstitium by the thick ascending limb of Henle's loop.

    d   Considerable calcium reabsorption occurs in the thick ascending segment.

    e   The thin ascending limb is impermeable to H_2O, highly permeable to NaCl, and moderately permeable to urea.

15 The distal tubule (DT):

    a   consists of a diluting segment followed by a concentrating one.

    b   reabsorbs more than 10% of the total sodium load.

    c   and the collecting duct are virtually impermeable to water when ADH is absent.

    d   is a site of secretion of K^+ ions in exchange for Na^+ ions.

    e   is the major site of renal NH_3 production.

16 Medullary hypertonicity:

    a   is essential if urine is to be concentrated.

    b   is primarily dependent on the variable permeability of the concentrating segment of the DCT to water.

    c   is partially maintained by delivery of water from ascending to descending limb of vasa recta (capillary loops).

    d   is partially maintained by the delivery of urea from the ascending to the descending vasa recta.

    e   is partially maintained by urea diffusing from capillary loops to the collecting ducts.

Tubule function (Answers)

10   a   **True.** Large amounts of water and solute are reabsorbed between the cells of the PCT.

    b   **False.** Two-thirds of the salt and water load is reabsorbed in the PCT.

    c   **True.** This Na^+–H^+ swop is counter-transport and is an important way of reabsorbing Na^+ in the PCT.

    d   **False.** Na^+, K^+-ATPase is present only on basolateral walls of the PCT epithelium.

    e   **True.** Increased hydrostatic pressure in the peritubular capillaries hinders reabsorption of water from the PCT.

11   a   **True.** Reabsorption rises if COP rises and falls if COP falls.

    b   **True.** As GFR increases, filtration fraction increases and increased COP in the peritubular capillaries assists in reabsorbing the extra fluid presented to the PCT.

    c   **False.** The tubular maximum (Tm) for glucose reabsorption is determined by the load of glucose presented per unit time to the PCT, i.e. plasma-glucose level × GFR.

    d   **False.** Glucose in PCT is cotransported with Na^+. Phlorizin blocks the symport protein carrier. However, it probably exits the cell to the inter-stitium via GLUT 2 (special glucose transporter).

e   **True.** 95% of PO_4^- is cotransported back with Na^+. PTH inhibits this. Phosphatonin probably stimulates PO_4^- reabsorption.

12 a   **False.** Movement of urea across membranes is always passive.

b   **False.** About 80% of K^+ is reabsorbed in the PCT. K^+ is both secreted and reabsorbed distally.

c   **False.** Most secretion occurs in the PCT. Secretion is often carrier-mediated.

d   **False.** Only ionized calcium passes through the glomerular membrane.

e   **True.** About 40% of filtered Ca^{2+} is reabsorbed by ascending limb of Henle's loop, DCT, and collecting ducts. PTH stimulates this fraction.

13 a   **True.** Urate is handled mainly in the PCT. About 10% of the filtered load appears in urine.

b   **True.** The satisfactory absorption of bicarbonate depends upon the presence of luminal H^+ ion and cellular carbonic anhydrase.

c   **False.** None the less, the cells of the DCT do not absorb much HCO_3^-.

d   **False.** Most H^+ ions are secreted in the PCT but are reabsorbed back with HCO_3^-.

e   **True.** Even histamine and acetylcholine are secreted by the PCT.

14 a   **True.** This is because it creates hypertonicity in the deepest medulla.

b   **True.** NaCl exits from the ascending limb to match H_2O that has diffused preferentially out of the descending limb.

c   **False.** Cl^- is actively expelled into interstitium and Na^+ follows. It was previously suggested that Na^+ was the 'primary' ion actively secreted.

d   **True.** Calcium and other cations replace some of the cations 'lost' as a result of the pumping out of Cl^- through the chloride channel.

e   **True.** The ascending and descending limbs have quite contrasting functions.

15 a   **True.** The first part of the DT is lined by epithelium similar to that in the adjacent thick segment of Henle's loop.

b   **False.** The DT reabsorbs about 60 mmol Na^+ or 5% of the total filtered Na^+ load.

c   **True.** ADH makes the second half of the DT and most of the collecting duct permeable to water by opening the water-channel proteins aquaporin-2 (in the luminal wall) and aquaporin-3 (in the basolateral wall).

d   **True.** $K^+/H^+/Na^+$ exchange is influenced by ion availability, urine flow rate, and aldosterone.

e   **False.** Most NH_3 is produced by the PCT, where it becomes NH_4^+ in the lumen. NH_4^+ substitutes for K^+ in the Na–K–2Cl system of the thick segment of loop, NH_4^+ is then reabsorbed and is later presented to the nearby collecting duct where NH_3 is secreted once more.

16 a **True.** Urine in collecting ducts achieves isosmotic equilibrium.

   b **False.** 25% of filtered NaCl is actively reabsorbed in the water-impermeable region of the thick segment of ascending limb. This switches on the counter-current concentrating mechanism and is the essential cause of hypertonicity in the medulla.

   c **False.** Water will flow osmotically into the ascending loop and so tend to maintain hypertonicity in blood going to the deepest medulla.

   d **True.** Any reabsorbed solute (including urea) that diffuses into the ascending loop passes into less concentrated blood of the descending loop .

   e **False.** Urea diffuses from collecting ducts into the interstitium and into the ascending capillary loops. This increases medullary hypertonicity.

Urea and tests of renal function (Qs 17–19)

17 *About urea:*

   a It is the chief nitrogenous waste product in man.

   b Normal people excrete about 5–8 g of urea in their urine per day.

   c Blood urea nitrogen (BUN) is greater than the blood urea.

   d Normally ureagenesis occurs almost exclusively in the liver.

   e If ureagenesis fails, then hyperammonaemia is inevitable.

18 *Blood urea levels (2.5–6.5 mmol/l):*

   a increase in the first week following parturition.

   b rise in patients with dehydration, e.g. from vomiting.

   c rise in patients with thyrotoxicosis.

   d commonly fall in patients with hepatitis.

   e rise in patients with bleeding into the GIT.

19 *Tests of renal function:*

   a Urine specific gravity (mass/volume) is a good index of renal concentrating ability.

   b The blood urea is a good test of renal function.

   c Serum creatinine is a better index of renal function than serum urea.

   d Serum uric acid is a better index of renal function than serum creatinine.

   e Beta 2–Microglobulin in plasma rises as renal function deteriorates, specifically as the GFR falls.

Urea and tests of renal function (Answers)

17 a **True.** Man and most terrestrial animals are ureotelic.

   b **False.** Humans excrete 20–30 g of urea in the urine per day (10–12 g N). Normal blood urea is 2.5–6.5 mmol/l.

   c **False.** Note: BUN is expressed as mg% (a rough guide is to divide blood urea value in mg by 2.5).

d **True.** Note: the first steps of ureagenesis take place in the mitochondrion.

e **True.** The urea cycle detoxifies NH_3 to urea and provides essential intermediaries for the citric acid cycle.

18 a **True.** Blood urea increases because of involution of uterine mass.

b **True.** Blood urea rises in dehydration because the same amount of urea must be accommodated in a smaller ECF volume.

c **True.** The blood urea rises in any hypercatabolic state.

d **False.** Liver function must be very depressed before urea falls.

e **True.** Uraemia of GIT bleeding is caused by absorption of excess protein and urea from the GIT. It is rather similar to the rise in blood urea that occurs when absorbing a large haematoma or when taking a high-protein diet. Blood urea also rises because of the action of urea-splitting bacteria on intraluminal blood.

19 a **False.** Specific gravity is affected by extraneous factors such as glucosuria and pyuria.

b **False.** Blood urea is also altered by diet, metabolism, and dehydration (see 18b, c, e).

c **True.** Serum creatinine is arguably a more useful test of renal function even than the creatinine clearance.

d **False.** Serum urate also depends on the extent of purine metabolism and on the ECF volume.

e **True.** Beta 2-Microglobulin is the light chain of the HLA antigen present on all nucleated cells. It is filtered by glomeruli and metabolized in the PCT.

Diuresis (Qs 20–21)

20 Diuresis is produced by:

a a rise in the GFR.

b hypoproteinaemia.

c excess fixed anion, e.g. the Cl^- part of NH_4Cl.

d osmotic diuretics because they dehydrate tubule cells.

e carbonic anhydrase inhibitors causing an alkalosis

21 Effects of specific diuretics:

a Loop diuretics, e.g. frusemide, block urine concentration by acting at the apex of Henle's loop.

b Loop diuretics can theoretically block the reabsorption of all the filtered load of sodium.

c The thiazide diuretics act proximally to the loop diuretics.

d Thiazides cause hypercalciuria.

e Potassium-sparing diuretics act distally on the $Na^+/K^+/H^+$ exchange.

Diuresis (Answers)

20 a **True.** GFR autoregulation can be overwhelmed, e.g. overtransfusion.

b **True.** In hypoproteinaemia there is a lower oncotic pressure and thus less water is reabsorbed from the tubules.

c **True.** Fixed anion robs luminal cation (Na^+,K^+) thus salt and water are lost in the urine.

d **False.** By being non-reabsorbable, osmotic diuretics (mannitol, sorbitol, sucrose) prevent H_2O being reabsorbed from the tubules.

e **False.** Carbonic anhydrase inhibitors, e.g. acetazolamide, cause acidosis and diuresis by blocking $NaHCO_3$ reabsorption (especially in PCT).

21 a **False.** Loop diuretics block active reabsorption of Cl^- in the thick segment of the ascending limb of Henle's loop.

b **False.** Theoretically loop diuretics can block about 25% of the filtered sodium load.

c **False.** Thiazides block Na^+ reabsorption in the first half of the DCT.

d **False.** Thiazides cause hypocalciuria. They are given to prevent nephrolithiasis in cases of hypercalciuria.

e **True.** K^+-sparing diuretics are weak diuretics and can only affect 2–5% of the filtered sodium load.

Renal failure (Qs 22–34)

22 *Acute renal failure (ARF):*

a is associated with oliguria (urine output < 20 ml/h) in over 90% of cases.

b is associated with a rise in plasma endothelin concentration.

c may be caused by free haemoglobin acting alone in the renal circulation.

d usually has a urine osmolality of less than 400 mosmol/kg.

e with a normal or greater than normal urinary output (non-oliguric ARF) indicates a predominance of tubular over glomerular damage.

23 *Plasma biochemistry changes in ARF:*

a Hyperkalaemia is common.

b Hypernatraemia is more likely in oliguric than in non-oliguric cases.

c Hypocalcaemia is more likely than hypercalcaemia.

d Hypophosphataemia is more likely than hyperphosphataemia.

e The rate of rise in blood urea is a good index of glomerular damage.

24 *When distinguishing prerenal failure from intrinsic renal failure:*

a A true urine specific gravity of 1.010–1.012 suggests intrinsic renal failure.

b Heavy albuminuria favours a diagnosis of renal failure.

c Urinary sodium is less in prerenal failure than in renal failure.

d There is little diuretic response to IV mannitol in either case.

e Urine/plasma osmolality ratio is at least 1.5:1.0 in functional renal failure.

25 The hepatorenal syndrome:

a occurs in patients with previously normal kidneys.

b can occur in the course of moderate liver failure.

c is usually associated with severe organic damage in both kidneys.

d Endothelin has been shown to be a major aetiological factor.

e Vasodilators such as nitric oxide may be aetiologically important.

26 Following nephrectomy:

a systemic hypertension develops eventually.

b hypertrophy of the remaining kidney occurs.

c acceptable fluid and electrolyte balance is maintained by as little as 10% of the original nephron population.

d there is an elevated glomerular filtration pressure in the remnant kidney.

e glomerular injury may by caused by elevated glomerular capillary pressure.

27 Chronic renal failure (CRF) or end stage renal disease (ESRD):

a is diagnosed once the creatinine clearance is consistently less than 50% normal.

b is commonly associated with hyperphosphataemia.

c is commonly associated with sympathetic overactivity.

d is consistently associated with raised plasma catecholamines.

e may show hypotension as often as hypertension.

28 Diet in chronic renal failure (CRF):

a Protein restriction may slow the progression of CRF.

b Protein restriction increases glomerular filtration.

c Protein restriction inhibits secretion of vascular transforming growth factor b.

d Fish-oil may slow the rate of deterioration in IgA nephropathy.

e Elimination of tyrosine and phenylalanine from the diet is nephro-protective.

29 Pharmacology of CRF:

a Long-term aspirin intake is consistently nephrotoxic.

b Long-term acetaminophen is consistently nephrotoxic.

c Phenacetin is relatively non-toxic to the kidneys.

d Phenytoin can cause a non-analgesic nephropathy.

e Nephrotoxicity from a specific drug is likely to be non-immune mediated if dose-related.

30 Uraemia:

a 'Uraemia' is clinically taken to mean CRF/ESRD.

b Oral calcium acetate can be used to reduce hyperphosphataemia.

c Secondary hyperparathyroidism is rarely seen in uraemia.

d Radiocontrast agents are unlikely to worsen CRF.

e Erythropoietin therapy may worsen hypertension in CRF.

31 With uraemia and CRF:

a In complete renal shut down or ESRD creatinine accumulates in the blood in proportion to its production.

b Plasma creatinine is an index of the severity of the renal failure.

c Insulin-dependent diabetics may require increased insulin dosage in uraemia.

d Urea is the major toxin in uraemia.

e Hyperkalaemia is not clinically significant until renal function becomes very seriously impaired.

32 Renal dialysis in CRF:

a Haemodialysis is efficient at removing inorganic phosphate.

b should be instituted if blood urea consistently > 25 mmol/l (150 mg%).

c Haemodialysis is more efficient than peritoneal dialysis in removing small molecules.

d Haemodialysis is less likely to cause abrupt CVS disturbance than peritoneal dialysis.

e Biocompatible membranes in the dialyser are more likely than other membranes to engender cytokines.

33 Prognosis in chronic renal disease depends on:

a the ability of mesangial cells to produce and release stretch-induced growth transforming factor (GTF).

b the function in remaining or unaffected parts of same kidney.

c age; e.g. NSAIDS are more nephrotoxic in younger than in older age groups.

d the degree of glomerular hypoperfusion.

e ethnic origin.

34 Renal failure:

a The haemolytic uraemic syndrome (HUS) is a form of CRF.

b HUS, preceded by diarrhea, is a major cause of ARF in the elderly.

c HUS is related to thrombotic thrombocytopenic purpura.

d Idiopathic nephrotic syndrome is almost always benign.

e Blood urea only rises moderately in the nephrotic syndrome.

Renal failure (Answers)

22 a **False.** Oliguria occurs in about 70% of cases of ARF.

   b **True.** The endothelium-derived endothelin-1 is usually elevated in ARF and in hypovolaemic shock. It is a powerful vasoconstrictor and promitogen for many tissues. (There are three known endothelins.)

   c **False.** Hb is a nephrotoxin only when associated with hypoxia and acidosis.

   d **True.** In functional oliguria osmolality is usually > 500 mosmol/kg.

   e **True.** Non-oliguric ARF reflects the inability of the tubules to reabsorb water while the glomeruli continue to function fairly well.

23 a **True.** Hyperkalaemia in ARF is due to an inability of the distal nephron to secrete K^+ properly.

   b **False.** Hypernatraemia may occur when there is relatively more H_2O than Na^+ lost in urine. It occurs most often in diuretic-induced ARF.

   c **True.** Hypocalcaemia is due to impaired tubular reabsorption of calcium and a decrease in renal-activated vitamin D.

   d **False.** The kidney is unable to excrete PO_4 normally. Also note that there is an inverse relation between plasma calcium and phosphate.

   e **True.** Note that blood urea can also rise rapidly in hypercatabolism even if the kidneys are normal.

24 a **True.** 1.010–1.020 is virtually the specific gravity of plasma. Isothenuria occurs when plasma and urine have the same specific gravity.

   b **True.** Heavy albuminuria suggests severe glomerular damage.

   c **True.** A urinary Na^+ of less than 20 mmol/l is common in prerenal failure.

   d **False.** A good response to diuretics can be expected in prerenal failure.

   e **True.** U/P osmolality ratio of nearly 5:1 is theoretically possible in functional oliguria.

25 a **True.** Pre-existing renal disease is unusual in the hepatorenal syndrome. Nor do the kidneys show any sequelae in survivors.

   b **True.** Although hepatorenal failure is more likely in the course of severe or chronic liver failure, where presumably there is an incomplete hepatic clearance of endotoxin.

   c **False.** The renal component of hepatorenal failure is believed to be a functional disorder of the kidneys. Intense vasoconstriction in the renal cortex is typical.

   d **False.** Endothelin-1 and -3 are elevated in hepatorenal failure. This is probably due to decreased renal disposal.

   e **True.** NO may cause abnormal vasodilation, hyperdynamic circulation, and renal failure. NO antagonists may be of therapeutic use.

26 a **False.** There is no change in BP after nephrectomy, provided the remaining kidney is normal.

b   **True.** After a nephrectomy there is also a functional improvement in the remaining kidney.

c   **True.** The normal redundancy of renal function is not understood.

d   **True.** The elevated glomerular pressure in remnant kidney has traditionally been held to prolong life, although persistent elevation of this pressure eventually damages renal function.

e   **True.** Antihypertensive therapy and dietary protein restriction help to lower glomerular pressure to safe levels.

27 a   **False.** In ESRD C_{cr} falls below 12–15 ml/min (normal 130 ml/min).

b   **True.** Hyperphosphataemia is common but not invariable in ESRD. If hypophosphataemia occurs, it rarely does so before renal function has declined to about 25% normal.

c   **True.** Afferent stimuli from a failing kidney appear to stimulate the sympathoadrenal system.

d   **False.** Elevated amines in ERSD are probably due to the confounding effects of uraemia on prejunctional amine release or its plasma-clearance.

e   **False.** Hypertension occurs in about 80% of patients with ESRD.

28 a   **True.** However, the beneficial effect of protein restriction is often minimal in humans.

b   **False.** Protein restriction in CRF causes constriction of dilated afferent arterioles and thus decreases individual glomerular filtration pressure.

c   **True.** Vascular transforming growth factor b is a stretch-released factor that stimulates collagen deposition.

d   **True.** Fish-oil may help by lessening the production of inflammatory cytokines and of various lipid mediators so typical of IgA-induced nephropathy.

e   **False.** Diets designed to reduce antigen intake, e.g. gluten-free diets, are not much use in slowing the progress of CRF.

29 a   **False.** Evidence that aspirin is nephrotoxic is scanty and inconsistent.

b   **True.** Acetaminophen probably doubles the risk. It is concentrated in the renal papilla.

c   **False.** Phenacetin is highly toxic. It was withdrawn from the market in the 1950s. Acetaminophen is a metabolite.

d   **False.** Phenytoin (depresses IgA production) may even be nephroprotective.

e   **True.** Drugs are nephrotoxic via dose-related effects (tubular cytotoxicity, vasoconstriction) or, if non-dose-related, immunogenic damage should be suspected.

30 a   **True.** Uraemia is the clinical syndrome characteristic of CRF/ESRD.

b   **True.** Calcium acetate inhibits absorption of both Ca^{2+} and PO_4^-.

c   **False.** Phosphate loading reduces renal synthesis of active vitamin D and lowers serum calcium, thus leading to secondary hyperparathyroidism.

d **False.** Contrast agents cause medullary ischaemia and may precipitate acute renal failure.

e **True.** By improving anaemia, erythropoietin aggravates rheology problems of the blood.

31 a **False.** Some creatinine passes into the gut; some recycles with muscle creatine.

b **True.** Note, however, that creatinine is not the 'uraemic toxin'.

c **False.** Insulin is degraded in normal kidneys, probably less is required in uraemic patients.

d **False.** Most hold that urea is a minor uraemic toxin although 'uraemic' symptoms have usually appeared when the blood urea reaches 20 mmol/l (120 mg%) in CRF.

e **True.** Usually creatinine clearance is below 10 ml/min before the plasma K^+ rises above 6 mmol/l. Note the possibility of precipitating dangerous hyperkalaemia in CRF in patients taking 'low-salt' and 'high-fruit' diets.

32 a **False.** Dialysis regularly removes only 8–10 mmol of phosphate per session.

b **False.** Institute haemodialysis when uraemia persistently exceeds 240 mmol/l (240 mg%).

c **True.** However, haemodialysis is less efficient in removing large particles than peritoneal dialysis.

d **False.** Abrupt blood or CVS disturbance is unlikely with peritoneal dialysis.

e **True.** Far more cytokine, leucocyte, and complement is induced by bio-incompatible membranes than by biocompatible ones.

33 a **True.** Glomerular hypertension stimulates the mesangial cells associated with the basement membrane of glomerular capillaries to produce GTF. This compounds glomerular sclerosis (see 33d).

b **True.** 10% of normal kidney can maintain adequate renal function.

c **False.** NSAIDS nephrotoxicity: doubled under 65 years, tenfold over 65 years.

d **True.** Initial hypertension-induced dilation of afferent arteriole maintains GFR. This is followed by mesangial-induced progressive glomerulosclerosis (collagen deposition). This reduces GFR and glomerular function deteriorates.

e **True.** Generally there is a poorer outlook in those of Afro-Caribbean ancestry, especially in the case of hypertensive CRF.

34 a **False.** The haemolytic uraemic syndrome (HUS) is characterized by ARF, thrombocytopenia, and anaemia.

b **False.** HUS preceded by diarrhea is a major cause of ARF in children.

c **True.** Episodes of ischaemia/infarction, especially in kidney, are found in thrombotic thrombocytopenic purpura.

d **True.** Only a small number of people with idiopathic nephrotic syndrome have focal segmental glomerulosclerosis and so go on to CRF.

e **False.** Blood urea falls in the nephrotic syndrome. Serum proteins are usually about 30–40 g/l, albuminuria may reach 30 g/day.

Urinary bladder and calculi (Qs 35–38)

35 *In the urinary bladder:*

a transitional lining epithelial cells have a very high turnover rate.

b at rest the pressure in the posterior urethra > intravesical pressure.

c the urethral closing pressure (UCP) exceeds the intravesical pressure (IVP) during non-micturition-related straining.

d a negative UCP may occur in normal micturition.

e a negative UCP is associated with stress incontinence.

36 *Micturition:*

a The intravesical pressure only rises significantly once > 1 litre of urine is stored in the bladder.

b Even a few drops of urine in the posterior urethra excites afferent neural activity from the bladder.

c The sympathetic tonically inhibits the detrusor by direct inhibition of the parasympathetic in the bladder wall.

d Contraction of pelvic floor muscles is important for strong urine flow.

e Overflow incontinence only occurs if the bladder is abnormally large.

37 *Renal calculi:*

a In normal people the urine is often saturated or supersaturated with calcium salts.

b Most urinary calculi are radiopaque.

c Calcium phosphate is the most common type of renal stone.

d Hyperparathyroidism is the most common cause of a calcium-containing stone.

e Triple phosphate stones are typical of *E. coli* urinary infection.

38 *Renal calculi formation:*

a Renal stones are much more likely to occur in women than in men.

b A low calcium diet is important in preventing calcium stone recurrence.

c A lesser rate of calcium oxalate stones may be achieved if the subject has a high intake of calcium.

d Increased incidence can be expected with a high intake of animal protein.

e Citrate is useful in reducing the frequency of renal stones.

Urinary bladder and calculi (Answers)

35   a   **False.** Turnover rate of transitional epithelium is 100–250 days. This is very slow compared to the skin and the GIT (5–10 days).

   b   **True.** Urethral closing pressure (UCP) should be up to 50 cm H_2O.

   c   **True.** Although the UCP may now be less positive relative to intravesical pressure when straining at other tasks.

   d   **True.** A negative urethral closing pressure helps initiate normal micturition in both sexes.

   e   **True.** UCP is barely greater than intravesical pressure at rest because of lack of vesicourethral support in multipara. Thus, any rise in intra-abdominal pressure may initiate urination (stress incontinence).

36   a   **False.** Normally pressure rises steeply after the bladder has been distended by 500 ml of fluid. At lower volumes there is stretch-relaxation.

   b   **False.** Urine in the posterior urethra excites neural afferent activity from the local urethral mucosal receptors.

   c   **True.** The sympathetic also works via a direct beta-adrenergic effect on detrusor myocytes.

   d   **False.** Contraction of pelvic floor muscles, and especially of the external urethral sphincter, is needed to maintain a high outlet resistance and thus guards urinary continence.

   e   **False.** Overflow incontinence can occur with a very small bladder, e.g. the thimble bladder of tuberculosis or chronic cystitis.

37   a   **True.** Urine is saturated with calcium salts despite the fact that about 98% of filtered calcium is reabsorbed.

   b   **True.** They are radiopaque because 90% of renal stones contain calcium.

   c   **False.** Calcium oxalate accounts for > 75% of renal stones.

   d   **False.** Hyperparathyroidism is responsible for 5–7% of renal calculi.

   e   **False.** Phosphate stones are found in alkaline urine, especially if urine is infected with urea-splitting organisms such as *Proteus* and some staphylococci .

38   a   **False.** Up to 10% of men, compared with 3% of women, have stones during their adult life.

   b   **False.** Research has shown no benefit in the time-honoured advice of taking a diet low in calcium.

   c   **True.** Insoluble calcium oxalate precipitates in, and is lost from, the gut.

   d   **True.** This is presumably because of the calciuric effect of the consequent acid load engendered by a high-protein diet .

   e   **True.** Urinary citrate results in the excretion of highly soluble calcium citrate which mops up urinary Ca^{2+}. However, it is often inconvenient to take the amounts of citrate required.

5

Cardiovascular

Haemodynamics (Q1–10)

1 In Laplace's Law, T = Pr or P = T/r:

a r is the radius (thickness, width) of the vessel wall.

b   As written above it could apply to the aorta.

c   As written above it could apply to capillaries.

d P is the transmural pressure.

e P is normally very close to the intraluminal pressure in the aorta.

2 Using the law of Laplace it can be shown that:

a   the tension in the wall of the aorta is about 12 000 times greater than that in an average systemic capillary.

b   in any two vessels the wall tension will be equal if the transmural pressure and wall thickness are the same in both.

c   in long-standing hypertension large vessels are liable to rupture more readily than smaller ones.

d   small aneurysms are more liable to rupture than larger ones.

e   intestinal rupture is most likely in the caecum.

3 Haemodynamics:

a   Velocity of flow increases as cross-sectional area increases.

b   Velocity is inversely proportional to volume of flow.

c   The Bernoulli principle states that the sum of the kinetic energy and the pressure energy is constant.

d   Lateral or distending pressure falls in narrowed segments of a vessel.

e   Blood is a true Newtonian fluid

4 Regarding blood flow:

a   Poiseuille's law is not strictly applicable in the human CVS.

b   In Poiseuille's law, flow (Q) varies directly with the radius (r) of the tube.

c   Laminar flow requires a greater driving pressure than turbulent flow.

d   The Reynolds number (Nr) for blood represents the number of particles dissolved in blood.

e   A low Nr is associated with laminar flow, a high Nr with turbulence.

5 The ratio of cells to plasma (packed cell volume (PCV) or haemocrit):

a   is normally about 60%.

b   is the single most important determinant of blood viscosity.

c   falls rapidly during a haemorrhage.

d   rises when the subject is in a cold environment.

e   rises during prolonged vomiting.

6 *In terms of vascular resistance to blood flow (R):*

a Arterioles are the first site of resistance in the systemic circulation.
b Resistance to flow through a single capillary is greater than that through a single arteriole.
c R is more in a single vessel (diameter X) than in four equal, smaller, parallel ones (each with a diameter of $X/4$).
d At rest arterioles convert flow to either non-pulsatile or nearly so.
e R can be expressed by the driving pressure divided by the flow rate.

7 *Regarding vascular resistance (R):*

a Total resistance (TR) in a parallel circuit is always greater than in any of its individual components.
b As pressure increases flow increases linearly.
c Systemic vascular resistance (SVR) is in the range 600–1100 dynes.
d The term impedance (I) only applies to flow in rigid tubes.
e R is about the same in a long as in a short vessel of the same width.

8 *In the control of vascular resistance:*

a angiotensin I is a powerful vasoconstrictor.
b renin is a reasonably strong vasoconstrictor.
c serotonin (5-hydroxytryptamine, 5-HT) is a strong vasoconstrictor.
d acetylcholine (ACh) causes vasodilation.
e ergot alkaloids constrict limb and digital vessels.

9 *As a determinant of cardiac performance:*

a the preload is the load that stretches the heart at the end of systole.
b the end diastolic pressure (EDP) is a more accurate estimate of preload than the end diastolic volume (EDV).
c the afterload includes the preload.
d the compliance (C) of the heart determines the ability of the ventricles to fill during diastole.
e left ventricular contractility is indexed by systolic pressure in the aorta if the left ventricular loading is held constant.

10 *The cardiac cycle:*

a It lasts about 0.6 s when heart rate is 75 bpm.
b Duration changes when the heart rate slows.
c Systole shortens at fast heart rates.
d Normally flow is at a standstill or sometimes a minor backflow occurs when pressure in the aorta exceeds that in the left ventricle just before the aortic valves shut.
e Most filling of the ventricles occurs in late diastole.

Haemodynamics (Answers)

1 a **False.** r is the radius of the lumen.

  b **False.** The law of Laplace should take account of the thickness (w) of the aortic wall: rewrite, $T = Pr/w$.

  c **True.** Capillary wall thickness can be ignored.

  d **True.** Transmural pressure = intraluminal–extravascular pressure.

  e **False.** The aorta is subject to widely varying intrathoracic and intra-abdominal pressures. Extravascular pressure is generally negligible elsewhere, e.g. in limbs.

2 a **True.** The high tension in the aortic wall is obvious when you convert mmHg into dynes/cm (wall tension).

  b **False.** The radius of the lumen remains the outstanding variable (see 1b). Note that, when standing quietly, capillary pressure in the feet may reach 100 mmHg, yet wall tension is only 1/3000th that of the aorta.

  c **False.** Arterial wall stress falls because of the thicker media.

  d **False.** An increasing r causes an increase in wall tension (T) and thus an increased tendency to rupture.

  e **True.** The large r of the caecum explains why intestinal rupture is more likely in the caecum than in narrower segments of the gut.

3 a **False.** Velocity is inversely proportional to cross-sectional area.

  b **False.** Velocity obeys the equation $V = Q/A$.

  c **True.** The Bernoulli principle means that the velocity of flow in a vessel is inversely proportional to the pressure distending its walls.

  d **True.** In a narrow segment, velocity of flow increases but the total energy remains virtually constant, thus the lateral distending pressure falls (Bernoulli principle).

  e **False.** Blood is a suspension, Newtonian fluids are homogeneous, e.g. water.

4 a **True.** Poiseuille's law applies strictly to steady laminar flow of Newtonian fluids in non-distensible cylinders.

  b **True.** Flow varies directly with the fourth power of the radius.

  c **False.** To produce a given flow the heart must work much harder if turbulence develops.

  d **False.** Nr represents the ratio of inertial to viscous forces.

  e **True.** For Nr < 2000 flow is usually laminar; if Nr > 3000 flow is usually turbulent.

5 a **False.** As usually expressed, the PCV is about 0.45 or 45%.

  b **True.** Plasma proteins normally contribute little to viscosity.

  c **False.** PCV falls only when there is a significant shift of tissue fluid into the vessels. Therefore a Hb count taken too soon will show a normal level despite significant blood loss.

    d **False.** However, cold *per se* does increase the viscosity of blood.

    e **True.** This occurs because the loss of ECF causes haemoconcentration.

6 a **False.** Aortic resistance is not negligible and comes first. Arterioles are the site of major resistance (variable size; small radius, 15 μm).

    b **True.** Yet there are so many capillaries that the pressure drop across them is not nearly as great as that across arterioles.

    c **False.** R is four times greater through the smaller parallel vessels, provided tube length and viscosity of the perfusate are held constant.

    d **True.** Arterioles convert flow into a non-pulsatile stream, except sometimes in exercise when they are widely dilated.

    e **True.** R in aorta = 90 mmHg (driving pressure) divided by 100 ml/s (flow), that is, R = 0.9 peripheral resistance units.

7 a **False.** It will be less since each parallel tube increases conductance ($C = 1/R$).

    b **False.** Flow and pressure have a linear relationship only in a non-distensible tube.

    c **True.** Note, however, that the pulmonary vascular resistance is ten times less (60–110 dynes) than the systemic resistance.

    d **False.** When applied to pulsatile flow, impedance incorporates resistance to mean flow rate as well as the vascular compliance that opposes the rate of change in flow.

    e **False.** Poiseuille's law also relates R to the length of the vessel. Note that the rate of an infusion depends on the length and width of the catheter and not on the size of the vessel used (provided other things, such as viscosity, are equal).

8 a **False.** Angiotensin II is a powerful vasoconstrictor.

    b **False.** In itself renin has little vasoactive power.

    c **False.** With intact normal endothelium 5-HT causes vasodilation via the release of endothelial nitric oxide.

    d **True.** ACh probably releases nitric oxide from intact endothelium (see 8c).

    e **True.** Note that gangrene can result from ergotism, e.g. from eating mouldy rye bread (infected with *Claviceps purpurea*).

9 a **False.** Preload is the stretching load just before systole at the end of diastole.

    b **False.** Preload (stretch) is better indexed by EDV. The EDP overestimates preload if compliance is reduced. EDP is more convenient to use clinically.

    c **True.** Afterload is the load that must be moved during contraction and so is part-set by the preload.

    d **True.** Compliance = end diastolic volume/end diastolic pressure.

    e **True.** Contractility is indexed by the systolic pressure achieved after the aortic valves open.

10 a **False.** Cardiac cycle lasts about 0.8 s (systole 0.25 s; diastole 0.55 s) at a heart rate of 75 bpm.

    b **True.** The slower the heart rate, the longer the cycle.

c **True.** Systole shortens at fast heart rate although diastole shortens far more.
d **False.** Antegrade flow continues due to inertia of blood.
e **False.** Most filling occurs in early or proto-diastole.

Heart sounds (Q 11)

11 Heart sounds:

a The first heart sound (S1) is due to closure of aortopulmonary valves.
b Normally the aortic valves close just before the pulmonary ones, causing splitting of the second heart sound.
c A third heart sound may be heard in normal children.
d Splitting of the heart sounds is always pathological.
e The fourth heart sound (S4) occurs during atrial systole.

Heart sound (Answer)

11 a **False.** S1 is due to closure of atrioventricular (AV) valves.
b **True.** Aortic valves close just before the pulmonary ones because of the larger pressure in the aorta than in the pulmonary artery.
c **True.** S3 is due to turbulence of flow in early diastole (rapid filling phase) and consequent vibration of the ventricular wall.
d **False.** Physiological splitting may be due to asynchronous contraction of the ventricles or increased filling of the right ventricle associated with ventilation.
e **True.** S4 is normally inaudible, it occurs as blood rapidly leaves the atria.

Cardiac Performance (Qs 12–16)

12 The Frank-Starling law:

a A Starling curve can be represented by relating the EDV to the systolic pressure.
b The steep ascending portion of the Starling curve represents the afterload.
c The descending limb of the curve may reflect overstretching.
d Hypervolaemia (increased end diastolic pressure, EDP) can decrease the cardiac output in man.
e Starling's law operates even when the cardiac volume is less than normal for a given individual.

13 Ventilation effects on cardiac performance:

a As the pleural pressure (Ppl) decreases, venous return increases.
b Increasingly negative Ppl increases blood flow from intra- to extrathoracic arteries.

   c The afterload on the left ventricle is increased by a rise in intraabdominal pressure.

   d There is a diversion of arterial blood towards the head during inspiration.

   e The tension in the ventricular wall during systole, that is the afterload, is lessened in inspiration.

14 Heart rate:

   a Maximum rate attainable stays much the same throughout adult life.

   b A very high rate reduces the stroke volume.

   c The heart rate slows considerably at rest if a normal subject takes a β blocker.

   d The heart rate slows considerably in exercise if a β blocker is taken.

   e A rapid IV infusion usually slows the rate if the initial rate is fast.

15 The central venous pressure CVP, ($2-10$ cmH_2O):

   a does not reflect a raised pressure in the left side of the heart.

   b increases if the compliance of the heart or of the pericardium is reduced.

   c remains the same throughout a respiratory cycle.

   d is a good index of the state of filling of the vasculature.

   e increases when there is constriction of capacitance veins.

16 In the jugular venous pulse (JVP):

   a The height is determined by the point of collapse of the internal jugular veins above the sternal angle.

   b Fine, rapid A waves occur in atrial fibrillation.

   c Large A waves (Cannon waves) occur in mitral stenosis.

   d Large V waves occur in tricuspid incompetence.

   e Hepato-jugulo reflux is a normal finding.

Cardiac Performance Answers

12   a **True.** Usually one puts Volume on the abscissa and Pressure on the ordinate. The clinical counterpart, the ventricular function curve, substitutes EDP for EDV and stroke volume for systolic pressure.

     b **False.** The steep ascending limb of Starling's curve reflects the preload and shows the importance of preload in augmenting contraction.

     c **True.** The descending limb is also a good index of the afterload.

     d **False.** The EDP must exceed 60 mmHg before ventricular output decreases. Such pressures are not encountered clinically.

     e **True.** Starling's law operates in the small heart, e.g. where the inotropic state is enhanced (exercise, excitement).

13   a **True.** Venous return increases as Ppl decreases until flow-limitation occurs (choke-point segment).

b **False.** A negative Ppl impedes blood flow from the left heart by increasing the left ventricular afterload. Transmural pressure rises.

c **True.** Any increase in pressure on extrathoracic arteries raises the left ventricular afterload.

d **True.** Blood is 'diverted' to the head in inspiration because of localized increased impedance to intra-abdominal flow from raised intra-abdominal pressure.

e **False.** Afterload (T) is increased by the increased ventricular volume (r) associated with inspiration (Laplace, $T = Pr/w$) and by an increased transmural pressure when pleural pressure becomes more negative.

14 a **False.** The maximum heart rate (HR) falls from around 200/min (young adult) to about 140/min in the elderly. The rule is 200 − (age).

b **True.** This is because a very rapid HR shortens the ventricular diastolic filling time.

c **False.** This is because sympathetic activity on heart is minimal under resting conditions.

d **True.** Sympathetic activity increases markedly in exercise.

e **True.** A rapid expansion of the vasculature causes bradycardia via the baroreceptor reflex and possibly via Bainbridge reflex.

15 a **False.** The CVP reflects chronic not acute pressure changes in the left side of the heart.

b **True.** Note there is a very high CVP in chronic constrictive pericarditis.

c **False.** The CVP increases in expiration and decreases in inspiration.

d **True.** The CVP is a good index of the state of filling of the vasculature, provided blood can freely enter a normally functioning heart.

e **True.** Sympathetic venoconstriction is prominent in shock and in heart failure.

16 a **True.** The height of the JVP is given at that point where the atmospheric pressure just exceeds the vein pressure. In normal people one may just discern the **A** and **V** waves but not the **C** wave.

b **False.** 'A' waves are absent because of the ineffectual atrial contractions of atrial fibrillation.

c **False.** Cannon waves occur in third-degree heart block when atria contract against closed tricuspid valves.

d **True.** Blood regurgitates from right ventricle to right atrium in tricuspid incompetence.

e **True.** The hepatojugulo reflux is greatly exaggerated when the heart cannot accommodate venous return. The test is done by pressing over the liver for 30–50 s.

Coronary circulation (Qs 17–19)

17 *In the coronary circulation:*

   a flow only occurs when the heart is relaxed (diastole).

   b flow is normal to low pressure chambers during systole.

   c the heart is provided with a very good oxygen reserve.

   d there is a very close correlation between myocardial oxygen consumption and the coronary blood flow.

   e the endogenous nucleoside adenosine is a coronary vasodilator.

18 *Concerning control of the coronary circulation:*

   a Coronary vessels are poorly supplied with sympathetic nerves.

   b Vagal nerve stimulation dilates coronary resistance vessels.

   c Hypoxia directly relaxes coronary resistance vessels.

   d A fall in pH relaxes coronary vessels significantly.

   e Glycerol trinitrate is a significant coronary vasodilator.

19 *Regarding vasoactive materials produced by the vascular endothelium:*

   a Prostacyclin (PGI_2) is a powerful vasodilator.

   b Production of endothelial PGI_2 is stimulated by pulsatile pressure.

   c The capacity to produce PGI_2 increases with age.

   d Nitric oxide synthase (NOS) is present in endothelial cells.

   e Nitric oxide and PGI_2 work via the same second messenger.

Coronary circulation (Answers)

17   a **False.** Flow occurs in systole to myocardium other than to that of the left ventricle.

      b **False.** This is because of occlusion of ostia by aortic valves during systole.

      c **False.** Despite a large blood flow, the heart has the highest oxygen extraction rate of any organ.

      d **True.** The relationship between myocardial oxygen usage and coronary blood flow holds true even in isolated or denervated heart.

      e **True.** Adenosine acts via purinergic receptors. It causes coronary vasodilation and slows the heart. Anoxia releases adenosine in both heart and brain.

18   a **False.** There is a very rich sympathetic supply to the coronary circulation.

      b **True.** The vagal dilatation effect is only of minor physiological significance.

      c **False.** Hypoxia is believed to vasodilate coronary vessels via the release of myocardial adenosine.

      d **False.** H^+, CO_2, or lactic acid appear to have little direct effect on the calibre of coronary blood vessels.

e **False.** The main 'cardiac' action of glycerol trinitrate is to reduce cardiac afterload by dilating non-cardiac vessels (via nitric oxide).

19 a **True.** PGI_2 is a vasodilator and an also a strong antiplatelet agent.

b **True.** PGI_2 is released as a result of pulsatile pressure and also by endogenous mediators (kinins, ADP, TxA_2, 5HT, thrombin).

c **False.** The production of PGI_2 decreases with age, diabetes, atherosclerosis, and cigarette smoking.

d **True.** Constitutive NOS is present in the vascular endothelium, especially of the heart and brain. Inducible synthase is found widely in the vascular endothelium and in macrophages.

e **False.** PGI_2 works via cAMP, NO works via cGMP.

Gravity effects and venous return (Qs 20–22)

20 Position affects blood flow:

a On standing, the venous pressure in the dorsum of the foot increases to over 100 mmHg in less than 2 min.

b As much as 50% of the blood volume can sequester in tissue spaces in prolonged standing in a hot climate.

c In the Trendelenburg position, the patient lies semirecumbent with pillows under the knees.

d Good autotransfusion is achieved by leg elevation in acute blood loss.

e Gravity aids venous return from above the heart if the subject is upright.

21 Gravity influences the CVS:

a There is a rise in pulmonary capillary pressure on lying down.

b The increase in central blood volume (CBV) in recumbency is a major cause of sudden infant death syndrome (SIDS) or cot death.

c Pressure falls in arteries throughout the body on standing up.

d Venomotor tone in capacitance vessels increases on standing up.

e Cardiac output falls sharply on standing up.

22 Venous return to the right heart:

a depends critically on the mean systemic filling pressure (MSFP).

b increases if arterioles dilate.

c increases if venules constrict.

d is obstructed when there are powerful voluntary diaphragmatic contractions.

e is decreased when pressure falls in the leg veins.

Gravity effects and venous return (Answers)

20 a **True.** 1 cm Blood = 0.77 mmHg (feet usually > 100 cm below heart).

  b **False.** Up to 20% of the blood volume may sequester in tissue spaces on standing in the heat. Heat causes venodilation.

  c **False.** This is Fowler's position. In Trendelenburg's position the subject lies flat and the legs are elevated.

  d **False.** Leg vessels are already vasoconstricted and are virtually empty in compensated hypovolaemic shock.

  e **True.** And gravity will impede venous return from dependent vessels.

21 a **True.** And this rise in pulmonary capillary pressure may be sufficient to precipitate orthopnea in patients with cardiac failure.

  b **False.** SIDS is probably linked with the prone position but not with the supine position, and so is not necessarily due to an increased central blood volume.

  c **False.** Arterial pressure falls in vessels above heart level when one stands up.

  d **True.** Venomotor tone helps maintain venous return. It is mediated by the sympathetic nervous system.

  e **True.** Cardiac output falls by up to 25% on standing up. This is compensated for rapidly by the baroreceptor response.

22 a **True.** At MSFP of 7 mmHg, coronary output = 5 litres/min; at MSFP of 10 mmHg, coronary output = 8 litres/min.

  b **True.** Arteriolar dilation increases the *vis a tergo* (power from behind).

  c **False.** Constriction of venules merely causes blood to sequester in the microcirculation.

  d **False.** Diaphragmatic contractions increase the venous return. Note: the muscle pump is taken to mean the compression of veins by contracting skeletal muscles especially in the limbs.

  e **True.** A good example is walking, which causes a fall in pressure in leg veins and aids venous return if the vasculature is normal. Pressure may rise if the subject has varicose veins.

Cardiac output (Qs 23–29)

23 *The following statements relating to the cardiac output (CO) are true:*

  a The CO is normally expressed as the output of both ventricles/min.

  b The cardiac index (CI) is the cardiac output divided by the pulse rate.

  c In severe exercise the cardiac output may increase tenfold.

  d The end diastolic volume (EDV) is normally about 145 ml at rest.

  e Myocardial contractile strength declines markedly in old age.

24 Cardiac pacemakers:

a   A sinoatrial node (SAN) is present in both atria.

b   If both vagus and sympathetic are silent, the SAN discharges at about 50/minute.

c   Purkinje cells have an intrinsic discharge rate less than that of atrio-ventricular node (AVN).

d   The resting membrane potential of the sinoatrial nodal cells is greater than that of other cardiac cells.

e   Prepotentials are only found in pacemaker cells.

25 Cardiac action potentials (APs):

a   The main initial upstroke in both pacemaker and non-pacemaker cells is due to the entry of Na^+ into the local cells.

b   The long plateau phase is due to the exit of calcium from the myocardial cell.

c   Tetrodotoxin (blocks fast Na^+ channels) has little effect on the AVN.

d   At rest the fast sodium channels are closed by m-gates.

e   During the AP, calcium channels in the sarcolemma are opened by movement of the membrane potential towards positivity.

26 Cardiac depolarization:

a   spreads radially from the SAN at a rate of about 1 m/s.

b   travels by a specialized route to the left atrium.

c   is subject to overdrive suppression from excess vagal activity on the SAN.

d   has a delay of almost 1.0 s at the AVN.

e   travels fastest along the Purkinje fibres.

27 Concerning calcium ions in the cardiac action:

a   About 50% of Ca^{2+} for excitation/contraction comes from the ECF.

b   Depolarization permits calcium ions to enter myoplasm via voltage-gated channels.

c   At rest there is up to a tenfold greater concentration of Ca^{2+} in the myoplasm than in the surrounding ECF.

d   Myoplasmic Ca^{2+} acts via cAMP (the second messenger).

e   Troponin C binds directly to calcium and initiates contraction.

28 Calcium in the cardiac cell:

a   Relaxation of the heart depends on the resequestration of calcium.

b   Relaxation uses more energy than contraction.

c   Phosphodiesterase reduces the effectiveness of calcium as an inotropic agent.

d   Digoxin raises myoplasmic calcium ions by directly increasing cAMP.

e   Both inotropic and active relaxant effects of substances such as caffeine, nicotine, tyramine, and cocaine are due to mobilization of calcium in the cell.

29 *In cardiac dysrhythmias:*

a Adenosine is often successful in terminating atrial fibrillation.

b Somatostatin can be expected to slow the heart in some tachycardias.

c Atrioventricular nodal re-entrant (AVNRT) is a rare form of paroxysmal supraventricular tachycardia.

d An ectopic focus can cause a premature ventricular contraction (PVC).

e Simple asymptomatic atrial fibrillation impairs cardiac performance in exercise.

Cardiac output (Answers)

23 a **False.** The cardiac output is normally expressed as the output of each ventricle per minute, which should be virtually equal.

b **False.** CI = cardiac output per square metre of body surface (about 2.8 litres/m/m^2).

c **False.** The cardiac output increases about five times (from 5 litres/min to 25 litres/min) in exercise.

d **True.** The EDV is an important determinant of cardiac output.

e **False.** Myocardial contractile strength is preserved, but the relaxation process slows markedly with ageing.

24 a **False.** The SAN (normal cardiac pacemaker) is only in the right atrium.

b **False.** The intrinsic rate of discharge of the SAN is about 100/min.

c **True.** Purkinje fibres fire 20–40/min; AVN fires 40–60/min.

d **False.** Transmembrane potential of SAN is –50 mV; it is –90 mV in other cardiac cells.

e **True.** Prepotentials are almost certainly due to the entry of Na$^+$ into the cell and/or the declining outward passage of K$^+$.

25 a **False.** Initial upstroke is probably due to Ca^{2+} entry in pacemaker cells and to Na$^+$ entry in non-pacemaker cells.

b **False.** The plateau phase of the AP is due to the entry of Ca^{2+} (L-type channels) and of some Na$^+$ via slow sodium channels.

c **True.** Calcium blockers decrease amplitude and duration of APs at the AVN.

d **True.** Any process that depolarizes the resting membrane potential tends to open the m-gates and allows Na$^+$ to enter.

e **True.** Long-lasting (L-type) calcium channels are abundant in the heart and are activated by AP upstroke. T-type (transient) channels are less common and are refractory to calcium-blocking drugs.

26 a **True.** Three specialized bands are the principal routes to the AVN from the SAN.

b **True.** Bachman's bundle carries the impulse from the SAN to the left atrium .

c **False.** Overdrive suppression refers to the suppression of latent pacemakers by greater intrinsic rhythmicity of the SAN.

d **False.** AVN 'nodal delay' is about 0.25 s. This co-ordinates atrio-ventricular contractions and makes them effective.

e **True.** APs in Purkinje fibres are similar to those in general ventricular fibres (they conduct rapidly and have a long refractory period).

27 a **False.** Most Ca^{2+} comes from stores in the sarcoplasmic reticulum (SR).

b **True.** Also the small amount of Ca^{2+} that enters releases SR Ca^{2+}.

c **False.** A 10 000-fold gradient exists between Ca^{2+} in the myoplasm and in the ECF. Ca^{2+} is extruded from myoplasm in exchange for Na^+ and by an ATP-dependent Ca^{2+}/H^+ pump.

d **True.** cAMP acts on protein kinases to phosphorylate cellular proteins.

e **True.** Phosphorylation of the troponin–tropomyosin complex decreases the affinity of the calcium receptor site on troponin C.

28 a **True.** Calcium must be removed from contractile filaments before relaxation occurs.

b **True.** Relaxation (lusitropy) is more energy consuming than contraction and is therefore more vulnerable to anoxia and ischaemia.

c **True.** Milrinone inhibits phosphodiesterase. This is why milrinone is an inotrope. Note: phosphodiesterase breaks down cAMP.

d **False.** Digoxin raises calcium by blocking the sarcolemmal Na^+/K^+ ATPase pump.

e **True.** The substances mentioned effect their action both directly via calcium and indirectly via sympathetic (increases cAMP).

29 a **False.** Adenosine (and verapamil) block the AVN. They are ideal for use in supraventricular tachycardias originating in the AVN.

b **True.** Somatostatin (SS) may slow the heart by reducing calcium entry to cells. SS is present in vagal cardiac fibres.

c **False.** AVNRT is the most common cause of this dysrhythm. Surgical interruption of pathways between the AVN and the atria often terminates the dysrhythmia.

d **True.** An ectopic focus can cause a PVC provided the timing is right. Clinically a 'dropped beat' is found.

e **True.** Atrial contraction is important in exercise when ordinary diastolic filling time is reduced ('atrial booster').

Autonomic controls (Qs 30–34)

30 Autonomic control of the CVS:

a In the heart, transmitters from the sympathetic (noradrenaline, NA) and vagus (acetylcholine, ACh) inhibit each other's release from nearby nerve terminals.

b Some parts of the heart are more susceptible than others to NA or ACh.

c In rapid eye movement (REM) sleep there is an increase in sympathetic discharge to the CVS.

d Carotid and aortic baroreceptors are normally most sensitive below 80 mmHg.

e Baroreceptors act as barostats.

31 In regard to the systemic baroreceptors:

a a neck-chamber pressure device is useful for studying the aortic baroreflex.

b when stretched, they show a reduction in afferent neural traffic.

c they are situated in the endothelium of the aortic arch and carotid sinus.

d activation stimulates the cardio-inhibitory area.

e the impaired baroreceptor response in ageing is especially marked on the heart rather than on the extracardiac vessels.

32 Fainting (vaso-vagal attack):

a typically shows a slowing of the pulse before the a loss of consciousness.

b is the most common form of syncope over 50 years of age.

c can only take place when the subject is upright.

d is accompanied by a massive vasodilation in skeletal muscles.

e can be successfully reversed by giving atropine to restore the blood pressure.

33 In syncopal and drop attacks:

a Orthostatic hypotension is always pathological.

b Syncope may be caused by a sensitive mechanoreflex from the left ventricle.

c Fainting in severe haemorrhage may be initiated by the ventricular mechanoreflex despite adequate cerebral perfusion at that time.

d Morphine or other opioids are useful in blocking the sympathetic decompensation of severe haemorrhage.

e Vasoactive amounts of arginine vasopressin (AVP) are released at the onset of syncope.

34 Ventilation affects heart rate:

a Sinus arrhythmia is a slight slowing of the heart during inspiration and a slight acceleration during expiration.

b Sinus arrhythmia is often exaggerated after exercise.

c Episodic falls in heart rate after moderate exercise is a normal phenomenon.

d Heart rate changes during the Valsalva manoeuvre are mainly mediated by the cardiac vagi.

e The Valsalva manoeuvre is normally followed by a tachycardia.

Autonomic controls (Answers)

30 a **True.** Mutual local inhibition between sympathetic and parasympathetic may be due to neuropeptides from both types of terminal.

  b **True.** Vagus dominates SAN, sympathetic dominates ventricles.

  c **True.** Thus rapid eye movement (REM) sleep, with enhanced sympathetic discharge, is a period of potential CVS risk.

  d **False.** The baroreceptors are most sensitive within the normal physiological range of blood pressure.

  e **False.** On the contrary, the baroreceptors readjust within minutes to the direction in which the blood pressure is moving.

31 a **False.** A neck-chamber pressure device is useful in the study of carotid but not intrathoracic baroreceptors.

  b **False.** Afferent neural traffic increases on stretching the baroreceptors.

  c **False.** Stretch sensors are situated mainly in the muscular and adventitial layers

  d **True.** Central action on the cardio-inhibitory centre is probably mediated via the local release of glutamic acid.

  e **True.** Impaired cardiac response is seen in old age, atheroma, exercise, and autonomic neuropathies. It manifests clinically as postural hypotension, causing faints.

32 a **False.** In a simple faint the pulse rate increases before one loses consciousness.

  b **False.** A simple faint is the most common cause of syncope in the young. Organic causes predominate in later years.

  c **True.** Thus a postural factor, e.g. lowered CVP, is important in the genesis of a simple faint.

  d **True.** Low levels of adrenaline cause vasodilation in skeletal muscle via beta receptors. Occupied cholinergic receptors also cause vasodilation in skeletal muscle.

  e **False.** Because syncope is primarily due to inhibition of the sympathetic to heart and vessels rather than to vagal-mediated cardiac slowing.

33 a **False.** Orthostatic hypotension may occur if one assumes the upright position quickly after having laid down in a hot environment such as a bath.

  b **True.** 'Ventricular syncope' is mediated by cardiac afferents carried along with the vagus.

  c **True.** Vigorous contractions of heart stimulate 'vagal' afferents which block sympathetic activity centrally.

  d **False.** Naloxone (an opioid antagonist) is useful because decompensation appears to be mediated by the release of opioids in the brain stem.

  e **True.** The non-osmotic release of AVP is due to intense baroreceptor unloading at the time of fainting.

34 a **False.** The opposite occurs. Vagal tone decreases during inspiration, thus the heart speeds up at this time.

    b **True.** Sinus arrhythmia is common after exercise in healthy young adults. It is due to persistent hyperventilation after the exercise has ceased.

    c **True.** Post-exercise vagushalt is uncommon but normal. Occurs at a time when vagal tone is trying to be restored to the heart.

    d **True.** The Valsalva test is often abnormal in diabetic autonomic neuropathy.

    e **False.** Post-Valsalva bradycardia is typical. It is mediated by the baroreceptors which respond to the steep rise in blood pressure as the restored cardiac output enters a tightly constricted arterial tree.

E.C.G. (Qs 35–36)

35 In the electrocardiogram (ECG):

    a The PR interval extends from the peak of the P wave to the peak of the QRS wave.

    b Ventricular depolarization proceeds from endocardium to epicardium.

    c An inverted P wave in leads 1 and 11 can be due to dextrocardia.

    d The normal QRS complex lasts about 0.2 s.

    e The limits of a normal QRS axis is $-30°$ to $+90°$.

36 On an abnormal electrocardiogram:

    a A Wenckebach phenomenon is likely to be seen in the sick sinus syndrome.

    b In atrial fibrillation the SA node fires up to 600/min.

    c Hyperkalaemia of 6–7 mmol/l causes changes in the T waves.

    d Widespread ST elevation is normally the first ECG change after a myocardial infarction.

    e ST depression of over 1 mm after an exercise test is abnormal.

E.C.G. (Answers)

35 a **False.** The PR interval extends from the beginning of the P wave to the beginning of the QRS wave.

    b **True.** It repolarizes in the opposite direction to depolarization, thus giving an upright T wave.

    c **True.** Dextrocardia or transposition of right and left arm leads can give rise to inverted P waves in leads 1 and 11.

    d **False.** The QRS complex should not exceed 0.10 s in any lead.

    e **True.** A more negative QRS axis is found in left-axis deviation, a more positive one in right-axis deviation.

36 a **False.** The Wenckebach phenomenon is a progressive increase in the PR interval until a P wave is not conducted, and is typical of damage to the AV node (Mobitz type 1 block).

b **False.** The AV node is bombarded by impulses at about 600/min from all over the atria in atrial fibrillation.

c **True.** In hyperkalaemia T waves in precordial leads become tall, slender, and peaked.

d **True.** ST elevation precedes the appearance of Q (necrosis) waves although T-wave inversion may occur first.

e **True.** ST depression of over 1 mm is strongly suggestive of ischaemic heart disease on exercise testing.

Oedema, heart failure, hypertension and age (Qs 37–41)

37 Oedema:

a Tissue pressure is relatively unimportant in controlling oedema.
b Oedema automatically occurs if pressure in major arteries rises.
c Tissue fluid normally has no oncotic pressure.
d Oncotic pressure of plasma falls on recumbency.
e Lymphatics normally return about 4 litres of excess tissue fluid/day.

38 In chronic heart failure (CHF):

a the sympathoadrenal system is always activated.
b renin release is generally depressed.
c plasma atrial natriuretic factor (ANF) rises.
d oedema is mainly due to increased venous pressure.
e diastolic dysfunction is more common than systolic dysfunction.

39 When under strain in the myocardium:

a myocardial cells respond by accelerating abnormal protein synthesis.
b left ventricular hypertrophy (LVH) is related to the height of the aortic pressure.
c damage will always ensue once chronic ventricular hypertrophy occurs.
d wall stress always increases in LVH (Laplace relationship).
e humoral agents can cause ventricular hypertrophy.

40 Essential hypertension:

a A mishandling of sodium is found in most patients with essential hypertension.
b The hypertensive effect of Na^+ is worse when it is present with the Cl^- anion.

c Weight reduction often restores normotension even in the non-obese hypertensives.

d Eskimos rarely suffer from essential hypertension.

e Increased intake of potassium helps prevent the onset of essential hypertension.

41 Age-related changes in CVS mean that:

a the ability of the heart to hypertrophy diminishes.

b the number of SA node cells remains more or less constant.

c degenerative changes are much more marked in the left than in the right coronary artery.

d the end diastolic pressure (EDP) in all cardiac chambers is higher in middle-aged and elderly than in young adults.

e postprandial hypertension is more common.

Oedema, heart failure, hypertension and age (Answers)

37 a **False.** For example, bandaging, which raises tissue fluid pressure, can control limb oedema.

b **False.** Arteriolar constriction protects capillaries from any rise in pressure in major arteries.

c **False.** Tissue fluid always contains at least a small quantity of proteins.

d **True.** There is reabsorption of hypo-osmotic ECF from dependent areas when the subject lies down.

e **True.** Note also that lymphatic oedema is non-pitting because of a high protein content, including clotting factors.

38 a **True.** Persistent activation of the sympathoadrenal system helps improve failing cardiac function despite agonist-induced down-regulation of β receptors in chronic heart failure. Note: cardiac $\beta2$ receptors are more potent at activating intracellular signalling than the predominantly occurring $\beta1$ receptors.

b **False.** Renin output rises in response to increased renal sympathetic activity and decreased renal perfusion pressure.

c **True.** Stretched atrial cardiocytes release ANF.

d **False.** Cardiac oedema is more likely due to aldosterone (enhanced sensitivity or larger amounts secreted or decreased hepatic degradation).

e **True.** Diastolic heart failure is very common, especially in the elderly where there is decreased cardiac compliance.

39 a **True.** In response to overload, myocardial cells make abnormal membrane and contractile proteins (especially 'slow' myosin) and there is a reversion to fetal isoforms.

b **True.** A rough correlation exists between LVH and aortic pressure. LVH is defined as a left ventricular mass of > 100 g in women and > 131 g in men (both per square metre). Note: the right ventricle performs one-quarter the stroke work yet it has only one-sixth the muscle mass of the left ventricle.

c **False.** Physiological hypertrophy occurs in athletes. This may be due to an intermittent rather than a continuous rise in afterload.

d **False.** For example, in concentric hypertrophy the relationship between cavity volume and wall thickness is unaltered. However, in eccentric hypertrophy the cavity volume is disproportionately increased.

e **True.** Agents such as adrenaline and angiotensin can affect ventricular size.

40 a **True.** A mishandling of sodium is found in a majority, but not in all, cases of hypertension.

b **True.** $NaHCO_3$ or sodium citrate appear to be less harmful than NaCl. The reason is not understood.

c **False.** Weight reduction is only beneficial in obese hypertensives.

d **True.** The rarity of hypertension in Eskimos is assumed to be due to fish-oils, which lower blood lipids, inhibit platelet aggregation, and increase the PGI_2/TxA_2 ratio. However, haemorrhagic strokes are much commoner.

e **True.** Potassium salts are definitely protective. This may provoke persistent natriuresis. Some workers recommend the addition of 10–20 mmol KCl to the normal daily diet. Beware of hyperkalaemia in patients on NSAIDs with renal impairment or on K^+ sparing diuretics.

41 a **True.** This is probably due to the age-associated loss of myocytes.

b **False.** A 75-year-old has only 10% of the SA cells of a 20-year-old.

c **True.** Degenerative changes begin in early adulthood in the left coronary artery; changes in the right one appear in the fifties.

d **True.** EDPs are higher both at rest and in exercise in the elderly, due to decreased cardiac compliance.

e **False.** Postprandial hypotension in the elderly can lead to syncope or angina. Slight postprandial hypertension is a normal occurrence in younger people.

Shock (Qs 42–46)

42 Shock:

a The toxic shock syndrome only occurs in females.

b Toxic shock syndrome is typified by sudden multiple organ failure.

c An otherwise normal adult can sustain a rapid loss of 30% of blood volume.

d In haemorrhage, fluid from tissue spaces makes up about 25% of the overt blood loss.

e Isotonic crystalloid replacement is an excellent temporary treatment in haemorrhagic shock.

43 *In haemorrhagic shock:*

a   if blood is unavailable give 3 ml of isotonic fluid for every 1 ml of estimated blood loss.

b   hypertonic saline and hyperoncotic dextrans are superior to isotonic fluids.

c   always initiate fluid resuscitation as soon as possible.

d   patients on β-blockers have impaired response to haemorrhage.

e   with massive bleeding, vasoconstriction persists until death.

44 *In septic or 'Gram-negative shock':*

a   Gram-negative bacilli are always the cause, even if not always isolated.

b   the toxic lipid moiety (lipid A) typical of rough strains of *E. coli* is a potent cause of the symptoms.

c   non-septic insults can also result in this syndrome.

d   tumour necrosis factor (TNF) plays an important role.

e   peripheral vasoconstriction is present initially in most cases.

45 *About shock (miscellaneous aspects):*

a   TNF stimulates the release of interleukins in septic shock.

b   Decreased plasma nitric oxide is usually found in septic shock.

c   The CVP is always low in hypovolaemic shock.

d   A fall in arterial lactate is a good prognostic sign, provided lactate levels were initially raised.

e   There is a poor prognosis in any type of shock if oxygen consumption exceeds 120 ml/min/m^2.

46 *The Acute Physiology and Chronic Health Evaluation II (APACHE II) Score:*

a   is useful in monitoring burnt patients.

b   is the sum of Acute Physiology, Age, and Chronic Health scores.

c   consists of 12 objective indices about the patient.

d   should be obtained within the first 24 h of admission if it is to be really useful.

e   correlates well with the death rate.

Shock (Answers)

42   a   **False.** About 50% of cases of toxic shock syndrome are associated with menstruation, the rest with extragenital infections. Males are also affected.

b   **True.** Respiratory, renal and hepatic failure, diarrhea, rash, and bleeding are all typical of the toxic shock syndrome.

c   **True.** A rapid loss of 30% of the blood volume can be sustained in normal healthy young adults. This amount is halved in the elderly and in those with autonomic dysfunction.

d **True.** Autogenic capillary infill begins within a few minutes of a fall in BP.

e **True.** The ideal of replacing blood lost by blood should only be acted on when other fluids are inadequate to sustain the patient's needs.

43 a **True.** Note that both crystalloid and colloid infusions increase the cardiac output better than whole blood because of their low viscosity. Thus they are superior therapies when one is concerned primarily with increasing oxygen delivery to tissues, provided the Hb is not decreased inordinately.

b **True.** Hypertonic fluids and plasma expanders help preserve the intra-vascular volume and reduce the total volume of fluid necessary for resuscitation.

c **False.** Aggressive IV fluids may dislodge thrombi. There is a better survival in many cases if IV fluid is minimized and the hypotension is tolerated until surgical exploration has been performed.

d **True.** The sympathoadrenal response is critical in haemorrhagic shock.

e **False.** When about one-third of blood volume is lost, sympathetic discharge may cease due to signals from an 'empty' heart.

44 a **False.** Septic shock sometimes accompanies a Gram-positive bacteraemia.

b **True.** Lipid A appears to be the active agent of bacterial endotoxin.

c **True.** In many cases, e.g. in burns, trauma, the ischaemic liver cannot cope with bacterial assault from another body source, e.g. gut, urinary tract, skin.

d **True.** Endotoxin is a most powerful stimulus for TNF production.

e **False.** However, the initial hyperdynamic circulation of septic shock may or may not progress to a hypodynamic one.

45 a **True.** Interleukins show a marked rise in septic shock, especially IL-1 and TNF. Note that both TNF and IL-1 have similar biological effects.

b **False.** There is usually an increased NO synthesis in septic shock. Inhibitors of NO synthase, as well as inhibitors of TNF, are clinically useful.

c **False.** The CVP is initially low in hypovolaemic shock; once cardiac failure occurs the CVP rises.

d **True.** Note that arterial lactate may be normal or only slightly raised in septic shock despite the appalling mortality.

e **True.** Septic shock carries a poor prognosis if there is more than double the normal oxygen consumption.

46 a **False.** APACHE II does not apply to burn or to non septic postoperative coronary artery bypass graft (CABG) patients.

b **True.** In the APACHE system add up to 6 points for age and up to 5 points for chronic illnesses.

c **False.** The Glasgow Coma Scale (GCS), used as part of APACHE II, is not objective; the other indices are (temperature, blood gases, heart rate, BP, plasma electrolytes and creatinine, PCV, WBC, respiration rate).

d **True.** The APACHE score should be used on admission. Otherwise it is of little use. Note the GCS (for APACHE) is scored as 15 minus observed GCS.

e **True.** Mortality rises from 10% at a score of 10 to 95% at a score of 35+.

6

Respiration

Mechanics, volumes and capacities (Qs 1–8)

1 *During normal quiet tidal breathing:*

a   the dome of the diaphragm hardly moves at all.

b   the upper external intercostal muscles help aerate the bases of the lungs.

c   a paralysis of the external intercostal muscles reduces the tidal volume (TV) by about 10%.

d   there is a decrease in both intrapleural and intra-alveolar pressure during the inspiratory phase.

e   the actual amount of air shifted each time by an adult is about 250 ml.

2 *The intra-alveolar (IA) and intrapleural (IP) pressures:*

a   In normal quiet expiration the IP pressure remains subatmospheric.

b   In the upright position there is a marked difference in IP at the bases and apices of the lungs.

c   In normal people the IA pressure always exceeds the IP pressure.

d   Intra-alveolar pressure is similar to that in the dorsal oesophagus.

e   Alveolar distending pressure (ADP) is the IP plus the IA pressure.

3 *Concerning volumes and capacities:*

a   A capacity is two or more volumes added together.

b   The vital capacity (VC) can be used to monitor heart failure.

c   The VC falls significantly on recumbency in normal people.

d   The VC is a good screening test of lung function.

e   The residual volume (RV) is exhaled in maximal expiration.

4 *In measuring pulmonary function:*

a   The VC should be 50–55 ml/kg body weight.

b   During exercise the respiratory rate can rise to about 55/min.

c   Maximum breathing capacity (MBC) should be more than 10 times the resting minute volume (RMV).

d   Functional residual capacity is measured with a simple Benedict–Roth spirometer.

e   A normal person can fully expire in less than 1 s.

5 *The functional residual capacity (FRC):*

a   is related in a linear fashion to one's height.

b   falls markedly as one gets older.

c   is increased in emphysema.

d   rises in recumbency.

e   should normally be about 40% of the total lung capacity (TLC).

6 The relationship of fibre length in the diaphragm means:

a for the same contraction of the diaphragm there is a greater volume inspired if the diaphragm is at high level rather than at a low level in the chest.

b fibres in a high diaphragm are longer than those in a low one.

c during acute asthma there is severe shortening of diaphragm fibres.

d in chronic obstructive pulmonary disease (COPD) fibres are lengthened.

e the low diaphragm of COPD is harder to actively stretch than normal.

7 The closing volume (CV) of the lungs:

a is the volume of air expired before airways close off.

b is small (20–50 ml) during quiet expiration in normal people.

c is associated with the closure of airways in the apex of the lung before closure elsewhere.

d increases in recumbency.

e if it equals the tidal volume, is incompatible with life.

8 With persistent repeated chronic forced expiration, e.g chronic coughing:

a there is a more rapid transfer of gas across the alveolar/capillary membrane.

b the likelihood of rupturing alveoli is increased.

c there may be a blockage of flow in pulmonary capillaries.

d cor pulmonale is likely from direct pressure on the right heart.

e the work of breathing is increased enormously.

Mechanics, volumes and capacities (Answers)

1 a **False.** The diaphragm moves about 1 cm or so in quiet breathing and about 10 cm during hyperventilation.

b **True.** The upper external intercostal muscles raise the whole rib cage because of the relative fixation of the upper ribs.

c **False.** Paralysis of external intercostal muscles has no effect on TV because the diaphragm is so effective.

d **True.** This fall in pressure permits a favourable gradient for the entry of air into the lungs.

e **False.** The tidal volume is closer to 500 ml.

2 a **True.** The IP pressure is still subatmospheric in quiet expiration at about −1 or −2 cmH_2O (1 cmH_2O = 0.73 mmHg).

b **True.** The IP pressure is about 7.5 cmH_2O greater at bases due to the weight of the lung.

c **True.** This occurs even in forced expiration where the IA pressure may be +100 cmH_2O and IP pressure + 90 cmH_2O .

d **False.** Oesophageal pressure reflects IP pressure at the same level.

e **False.** The ADP is the transmural pressure, viz IA–IP pressure.

3 a **True.** For example, vital capacity = all volumes except residual volume.

 b **True.** This is because the VC varies with the fluid content of the lungs.

 c **True.** The increased blood in chest and rising up of the diaphragm reduce the VC in recumbency.

 d **False.** The VC mainly depends on chest compliance, muscle power, and body size.

 e **False.** The RV is the volume of air that we can never exhale from our lungs.

4 a **True.** However, in the neonate the VC is about 33 ml/kg body weight.

 b **True.** A respiratory rate of 55/min is about the maximum for elite athletes aged 20–30 years.

 c **True.** The RMV = 8 litres/min; the MBC = 110 litres/min in females and 130 litres/min in males (values are approximate).

 d **False.** The FRC = ERV + RV. RV cannot be measured by a simple spirometer.

 e **False.** Expiration takes about 3 s. After a full inspiration it may take up to 6 s to fully expire.

5 a **True.** The FRC is about 10% less in females than in males.

 b **False.** The FRC bears little relationship to age *per se.*

 c **True.** There is increased FRC in emphysema because of the loss of alveoli and the increase in alveolar size.

 d **False.** FRC falls in recumbency for the same reason that the total lung capacity falls (see 3c).

 e **True.** The FRC and TLC have a reasonably steady relationship independent of chest size.

6 a **False.** The cone shape of thorax means that a greater volume of air is drawn in if the diaphragm starts contracting from a lower position.

 b **True.** This may confer some contractile advantage (Starling's Law) on the longer fibres in a high diaphragm.

 c **True.** In acute hyperinflation the FRC doubles, fibres are passively shortened and the diaphragm loses virtually all its strength.

 d **False.** Fibres are chronically shortened (low diaphragm) in COPD. In such a case an adaptive loss of sarcomeres may restore power to the remaining ones in which length has been allowed to return to normal or near normal.

 e **True.** In COPD the diaphragm is passively shortened by its own elasticity. Thus there are more elastic forces to be overcome.

7 a **False.** The closing volume (CV) is the volume expired after the airways begin to close.

 b **False.** The CV only appears during maximal expiration.

 c **False.** Gravitationally dependent basal airways close off first.

 d **True.** CV increases because recumbency interferes with diaphragmatic movements. Obesity also increases the CV.

e **False.** When TV = CV the patient may be a respiratory cripple. Much of the lung is excluded from ventilation.

8 a **False.** Gas transfer may be slower because of raised extra-alveolar pressure.

b **False.** Persistent high extra-alveolar pressures cause many alveoli to close (unsupported segments of respiratory bronchioles may collapse).

c **True.** Repeated occlusion of pulmonary capillaries leads to pulmonary hypertension and cor pulmonale.

d **False.** Cardiac work is increased by repeated Valsalva-associated tachycardia and vasoconstriction.

e **True.** Death from respiratory muscle fatigue can occur in patients with persistent coughing, as in COPD or chronic bronchitis.

Dead space, $\dot{V}/\dot{Q}$, shunting and compliance (Qs 9–13)

9 Dead space (DS):

a The volume of the anatomical DS alters during a ventilatory cycle.

b At rest the anatomical and alveolar DS have roughly equal volumes.

c The alveolar DS becomes very large in exercise in normal people.

d One can measure the alveolar DS by O_2 analysis of expired air after inhaling 100% oxygen.

e The local alveolar DS decreases in a consolidated lobe.

10 Pulmonary ventilation:perfusion ($\dot{V}:\dot{Q}$) ratio:

a In the upright position the normal $\dot{V}:\dot{Q}$ ratio causes blood from the lung apex to be less than fully oxygenated.

b A decrease in the $\dot{V}:\dot{Q}$ ratio has the same effect as RL shunting.

c In severe pulmonary embolism shunting is greatly increased.

d Giving oxygen improves hypoxaemia in shunting states.

e Giving oxygen is more useful in a true shunt than in a shunt-like state, e.g. pulmonary trauma or collapse (atelectasis).

11 Gas diffusion in the lungs:

a Diffusion can be increased voluntarily in normal people.

b It is independent of the solubility of the gas in the alveolar capillary membrane.

c CO_2 diffuses better than O_2 within the alveolus.

d Cardiac action significantly aids in gas transfer in the lungs.

e Oxygen transfer is perfusion-dependent.

12 Regarding gas exchange in the lungs:

a The transit time of a red cell in the pulmonary circuit is about 1 s.

b It takes about 1 s for full diffusion of O_2 and CO_2 to occur.

c　CO combines about 25 times more avidly with Hb than O_2.

d　CO transfer across alveolar/capillary membrane is diffusion-limited.

e　The diffusion of CO_2 and O_2 across the alveolar membrane is proportional to the thickness of that membrane.

13 In the mechanics of breathing:

a　the compliance is given by volume/pressure (V/P).

b　the compliance of the lungs is about the same as that of the thoracic cage in young healthy adults.

c　the compliance of the thorax decreases in obesity.

d　the volume change lags behind pressure change in both inflation and deflation of the lungs.

e　an estimate of elastic forces is given by area between inspiratory and expiratory traces on the hysteresis loop.

Dead space, V̇/Q̇, shunting and compliance (Answers)

9　a　**True.** The anatomical DS is larger during inspiration.

b　**False.** Alveolar DS is normally less than 3% of alveolar ventilation.

c　**True.** The alveolar DS increases in exercise because hyperventilation increases the lung V̇:Q̇ inordinately.

d　**False.** This is Fowler's method for measuring anatomical dead space.

e　**True.** Local DS falls because the V̇:Q̇ falls in the lobe in question. Overall the V̇:Q̇ may rise because of hyperventilation.

10　a　**False.** Blood draining the lung apex is well oxygenated (V̇:Q̇ about 3:1).

b　**True.** Anatomical shunts (true shunts) involve bronchial vessels, the venae cordae minimae, Thebesian and anterior cardiac veins.

c　**False.** Pulmonary embolism causes increased V̇:Q̇ (increased alveolar dead space).

d　**True.** Giving oxygen is essentially similar to increasing the lowered V̇:Q̇.

e　**False.** In a true shunt the venous blood completely bypasses the ventilated lung. Thus O_2 inhalation is not of much use.

11　a　**True.** Hyperventilation increases the surface area of alveoli.

b　**False.** CO_2 is 25 times more soluble than O_2 in the alveolar–capillary membrane and so diffuses into the alveoli faster.

c　**False.** CO_2 is a heavier gas than O_2 and so less diffusible in a gaseous medium.

d　**True.** This is due to significant vibration of alveoli caused by cardiac movements.

e　**True.** Oxygen transfer ceases once the alveolar–capillary gradient is abolished (independent of its solubility in the membrane).

12 a **False.** An RBC takes about 1 s to traverse a pulmonary capillary. It moves faster in hyperdynamic states.

b **False.** Normally full diffusion and oxygenation takes about 0.3 s.

c **False.** CO combines with Hb about 210 times more avidly than O_2.

d **True.** Avidity of Hb for CO keeps plasma CO very low. The only obstacle to the transfer of CO from alveolus to Hb is its passage across the membrane. The CO diffusion test is low if the membrane is thickened, e.g. inflammation or in severe anaemia.

e **False.** Diffusion of gases is inversely proportional to the thickness of the membrane (see Fick's law of diffusion).

13 a **False.** Compliance is given by relating a change in V to a change in P.

b **True.** The compliances of the lungs and of the thorax in young adults are both about 0.2 litres/cm H_2O .

c **True.** Because obesity interferes with diaphragmatic descent.

d **True.** The lagging of volume change behind pressure change is the basis for the hysteresis loop.

e **False.** Area enclosed in the hysteresis loop can be used to estimate the non-elastic forces (expiration is passive).

The work of breathing (Qs 14–22)

14 Concerning the work of breathing:

a The oxygen cost of respiratory muscles increases proportionately as ventilation increases.

b Airway resistance is the main component in the work of breathing.

c Collagen contributes to the elastic resistance of the lungs.

d Elastin fibres are difficult to stretch.

e Over half the airway resistance is in the very small airways.

15 During the respiratory cycle:

a Inspiration occupies 40%, expiration 60% of a normal respiratory cycle.

b Paradox is present when inspiratory time exceeds expiratory time.

c Respiratory paradox is typically associated with pulsus paradoxus.

d Respiratory paradox can be caused by an obstruction in a large airway.

e In a patient with a flail chest the increase in abdominal volume in inspiration matches the amount of air drawn into the chest at that time.

16 The following are true of surface tension (ST):

a If the STs of two connecting bubbles are the same, then the smaller bubble will empty into the larger one.

b It is easier to expand an air-filled lung than a saline-filled lung.

c The surface tension of a flat water layer is independent of the area.

d The surface tension of water, after adding detergent, is dependent on the area of the water/air interface.

e The surface tension of water, after adding lung extract, is dependent on the area of the water/air interface.

17 Surfactant:

a is a glycoprotein.
b has both a rapid synthesis and turnover rate in the body.
c is produced by the fetus in the first trimester of pregnancy.
d is less effective in reducing surface tension at larger than at smaller lung volumes.
e is an important factor in preventing pulmonary oedema.

18 In infants suffering respiratory distress syndrome (RDS):

a there is a grossly reduced surfactant pool.
b it is more common in male than in female babies.
c expiration is a greater problem than inspiration.
d a respiratory rate over 60/minute is typical.
e it is not apparent clinically until at least 24 h after birth.

19 In the adult respiratory distress syndrome (ARDS):

a the lungs become diffusely stiff throughout.
b blood is shunted away from poorly ventilated regions.
c pulmonary oedema is mainly due to damage/loss of surfactant.
d pulmonary hypertension decreases blood flow to aerated regions.
e mechanical ventilation and PEEP (positive end expiratory pressure) may rupture normal alveoli. PEEP is used to prevent small airways closing at the end of expiration.

20 Airway resistance:

a is a major factor in the elastic resistance of the lungs.
b is mainly determined by the calibre of the airways
c falls at high altitude.
d is inversely related to the FEV1s (Forced expiratory volume in 1s).
e increases as lung volume increases.

21 Within the bronchial tree:

a Bronchoconstrictor vagal tone is greatest at night.
b Non-adrenergic non-cholinergic (NANC) nerves are bronchoconstrictor.
c The submucosal bronchial venous plexus extends almost to the alveoli.
d The mucociliary escalator is stimulated by the sympathetic system.
e Sinopulmonary infection, situs inversus, and male sterility can all be associated with ciliary dyskinesia.

22 In bronchial asthma:

a Expectorate in uncomplicated acute asthma has many neutrophils and eosinophils.

b there is a chronic progressive course with increasing fibrosis of the airways.

c Cells with IgE receptors (mast cells, macrophages) are crucial in the genesis of asthmatic symptoms in most patients with asthma.

d environmental factors can increase airway responsiveness.

e 'Cardiac asthma' results from bronchoconstriction associated with cardiac disease.

The work of breathing (Answers)

14 a **False.** Oxygen cost of breathing is about 0.5 ml/l in ventilation at rest, and about 4.0 ml/l in moderate hyperventilation.

b **False.** About 70% of the work of breathing is due to overcoming elastic forces.

c **True.** Elasticity is associated with the geometric arrangement of elastin and collagen fibres ('nylon stocking effect').

d **False.** Elastin stretches easily, collagen does not. Surface tension is also a very important elastic resistance.

e **False.** Only 2% of resistance occurs in the small airways (Nile delta effect); 50% in upper airways. Note the importance of gas diffusion rather than bulk flow in the small airways.

15 a **True.** Expiratory time in particular is shortened in hyperventilation.

b **False.** Relative lengths of phases have nothing to do with paradox.

c **False.** Pulsus paradoxus is an exaggeration of the normal fall in systolic BP during inspiration. It is independent of respiratory paradox.

d **True.** Exaggerated inspiratory effort creates an above-normal negative intrapleural pressure which draws in the lower ribs (Hoover's sign) and the intercostal spaces.

e **False.** The inspiratory increase in abdominal volume is much larger than the volume of air inspired because of the simultaneous depression of the flail segment of the chest.

16 a **True.** Pressure is greater in the small bubble ($P = ST/r$; Laplace).

b **False.** One needs a gas:liquid interface to get a ST effect.

c **True.** Surface tension of water is about 70 dynes/cm^2.

d **False.** Although detergent reduces the total surface tension to about 25 dynes/cm^2.

e **True.** The smaller the area the lower the ST (range 4–30 dynes/cm^2).

17 a **False.** Surfactant is 80% phospholipid, 8% neutral lipid, 12% protein. It is secreted by type II alveolar cells.

b **True.** Synthesis of surfactant is promoted by steroids, and to a lesser extent by thyroxine and barbiturates.

c **False.** Synthesis of surfactant only starts around the 24th–26th week. The viability of premature babies depends upon the commencement of surfactant synthesis.

d   **True.** The surface concentration of surfactant is decreased in larger alveoli.

e   **True.** Surfactant diminishes the leakage of fluid from capillary into alveolus.

18   a   **True.** In RDS the surfactant pool is reduced to 5 mg/kg (normal is about 100 mg/kg).

b   **True.** Oestrogen tends to promote fetal surfactant synthesis.

c   **False.** In RDS most trouble occurs in opening up collapsed alveoli. Persistent inspiratory rib retraction is a feature.

d   **True.** Severe hypoxaemia causes tachypnea greater than 60/min.

e   **False.** Virtually all cases of RDS become apparent within 4 h of birth.

19   a   **False.** Patchy distribution of areas with poor compliance and ventilation is typical of ARDS.

b   **True.** Hypoxic vasoconstriction helps divert blood to better aerated areas of the lung, e.g. at costophrenic angles. Some contend that this does not occur in ARDS. V/Q mismatch leads to hypoxaemia in ARDS.

c   **False.** Pulmonary oedema is due to a generalized increase in the permeability of the capillary membrane. Abnormalities of surfactant worsens any oedema and this is thought to be part of the pathogenesis.

d   **True.** Cytokines, eicosanoids, and complement cause platelet aggregation and vasoconstriction. Nitric oxide inhalation can help. Also, interstitial oedema causes compression of the lung vasculature.

e   **True.** Mechanical ventilation may rupture normal alveoli by causing hyperinflation in compliant regions of the lung.

20   a   **False.** Normally airway resistance comprises 80% of the non-elastic resistance.

b   **True.** Bronchomotor tone is an overriding control in airway resistance.

c   **True.** The work of breathing is reduced at altitude because inhaled air is less dense at altitude.

d   **True.** Expiration is especially affected by bronchial narrowing.

e   **False.** Airway resistance is inversely related to lung volume.

21   a   **True.** The diagnosis of asthma is in doubt if there are no dips in the early morning peak flow readings and if there are no nocturnal awakenings.

b   **False.** NANC nerves probably secrete vasoactive intestinal polypeptide (VIP), a bronchodilator.

c   **True.** The submucosal bronchial venous pool is important for heating and humidifying inhaled air.

d   **True.** Co-ordinated movements of the mucociliary escalator are inhibited by vagal stimulation.

e   **True.** This is Kartagener's syndrome, a genetic absence of dyenin (an ATPase that provides energy for ciliary movement). Note that without situs inversus, it is cystic fibrosis.

22   a   **False.** A unique feature of the expectorate in uncomplicated asthma is a lack of neutrophils and dominance of eosinophils.

b **False.** Asthma shows exacerbations and remissions, airways remain remarkably free of fibrosis.

c **True.** Mast cells and macrophages release chemical mediators that can be neurally amplified in IgE-dependent cases, including occupational and drug-induce asthma.

d **True.** Airway responsiveness is increased by air pollutants, dust, cigarette smoke, house dust mite, dry air, and exercise. (Note unloading of soyabeans in Barcelona and asthma outbreaks.)

e **False.** Cardiac asthma is a result of pulmonary and bronchial hypertension causing breathlessness.

Handling of blood gases (Qs 23–28)

23 Regarding the carriage of oxygen:

a The partial pressure of oxygen (PO_2) in the uppermost layers of the Atlantic ocean is far less than that in arterial blood (PaO_2).

b Normally the PaO_2 exceeds the $PaN2$.

c The total pressure of gases is the same in arterial and venous blood.

d The PaO_2 is altered if the concentration of Hb is altered.

e One gram of haemoglobin can combine with 1.39 ml oxygen.

24 Transportation of oxygen in blood:

a It is enhanced by an increase in the local PCO_2.

b The saturation of arterial blood with oxygen (SaO_2) reaches 100% when breathing pure O_2 at sea level.

c The PaO_2 rises very considerably if one inhales pure oxygen at sea level.

d The PaO_2 is likely to be normal in polycythemia vera.

e Oxygenated RBCs contain more water than those with reduced Hb.

25 When measuring blood gases:

a arterial blood samples are better than capillary ones for measuring PaO_2 in a clinical setting.

b arterial blood samples are significantly superior to mixed venous ones for assessing the efficacy of the lungs.

c oximetry distinguishes between oxyHb and deoxyHb.

d pulse oximeters differentiate Hb in arteries from Hb in veins.

e transcutaneous O_2 and CO_2 monitors depend on the actual diffusion of these gases across the epidermis.

26 The following statements are true about carbon dioxide in the blood:

a A low PO_2 favours increased carriage of CO_2 by Hb.

b About 10% of blood CO_2 is carried in the form of carbamino compounds.

c About 10% of blood CO_2 is carried dissolved in plasma as H_2CO_3.

d Plasma CO_2/O_2 = respiratory quotient of the subject.

e $PaCO_2$ is better than PaO_2 in defining the adequacy of ventilation.

27 *The following statements about plasma CO_2 are true:*

a The $PaCO_2$ normally bears a predictable relationship to alveolar CO_2 ($PACO_2$).

b The $PaCO_2$ (40 mmHg, 5.3 kPa, 1.2 mmol/l) is a good index of metabolic CO_2 production.

c A severe respiratory acidosis can occur with a normal $PaCO_2$.

d If the $PaCO_2$ is 40 mmHg (5.3 kPa), then the HCO_3^- should be 24 mmol/l.

e CPR in cardiac arrest lowers the $PvCO_2$.

28 *End-tidal CO_2 ($ETCO_2$):*

a This is best measured at the end of a forced or maximal expiration.

b This is closer to the $PvCO_2$ (mixed venous) than to the $PaCO_2$.

c A high $PaCO_2$–$ETCO_2$ gradient is found if the alveolar dead space increases.

d A high $PaCO_2$–$ETCO_2$ gradient is expected in low cardiac output states.

e A sudden rise in the $ETCO_2$ may occur in malignant hyperthermia.

Handling of blood gases (Answers)

23 a **False.** The PO_2 in the upper layers of the Atlantic is greater and far closer to that of the atmosphere than the PaO_2 of arterial blood.

b **False.** PaN_2 is 76.4 kPa or 573 mmHg; PaO_2 is 100 mmHg or 13.4 kPa.

c **False.** Total gas pressure is greater in arterial than in venous blood. Note that pressurewise far more O_2 is lost from arterial blood than CO_2 is added to venous blood.

d **False.** The PaO_2 depends on the amount of O_2 in solution in plasma, e.g. the PaO_2 is normal in anaemia.

e **True.** *In vivo* 1 g combines with 1.34 ml O_2 because of the Met- and CO-Hb already present.

24 a **False.** Increased PCO_2 favours O_2 dissociation from oxyHb (Bohr effect).

b **False.** 2% of pulmonary blood is normally shunted via bronchial circulation. SaO_2 will never reach 100% even if one breathes 100% oxygen.

c **True.** Alveolar PO_2 (PAO_2) increases as O_2 replaces N_2 when one breathes pure oxygen.

d **False.** In polycythemia vera the PaO_2 is lower than normal. There is insufficient time for oxygen to fully saturate the excess Hb.

e **False.** H_2O enters RBC along with CO_2 as oxyHb sheds O_2 in tissues.

25 a **False.** Samples from an arterialized capillary, e.g. warmed earlobe, are just as accurate as those from arteries.

b **True.** PvO_2 is no index of pulmonary function.

c **True.** Oximetry is based on the ability of different forms of Hb to absorb lights of different wavelengths (oxyHb, red; deoxyHb, infrared).

d **True.** Pulse oximeters measure light transmission through pulsatile vessels.

e **True.** Transcutaneous gas monitors are more accurate if the skin is warm and thin, e.g. in neonates.

26 a **True.** The ability of a low PO_2 to facilitate CO_2 carriage is known as the Haldane effect.

b **True.** The amount carried as carbamino compounds is fairly constant at around 30–40%. It is found mainly in the RBC with Hb.

c **False.** Less than 5% CO_2 is dissolved. H_2CO_3 forms slowly in the plasma, but quickly in the RBC (i.e. when carbonic anhydrase is present).

d **False.** The RQ is the CO_2/O_2 in expired air (about 200/250; 0.8).

e **True.** There are many non-pulmonary causes of hypoxaemia, e.g. anaemia, low O_2 tensions in air or heart failure.

27 a **True.** Normally about 5–8 mmHg difference between $PACO_2$ and $PaCO_2$.

b **False.** The $PaCO_2$ gives no index of tissue CO_2 when ventilation is normal.

c **True.** Circulatory failure, shock, and cardiac arrest prolong the capillary transit time and give a normal $PaCO_2$, an increased $PvCO_2$, and a decreased arterial (and venous) pH.

d **True.** PCO_2 of 40 mmHg = 1.2 mmol/l. $NaHCO_3/H_2CO_3$ = 20:1 at pH 7.4.

e **False.** CPR lowers the $PaCO_2$ but the $PvCO_2$ will rise because of tissue acidosis. Giving $NaHCO_3$ will cause a further rise in the $PvCO_2$ ($NaHCO_3$ generates CO_2 with tissue lactic acid).

28 a **False.** The $ETCO_2$ reflects the highest CO_2 level reached on a capnogram. It occurs at the end of a normal tidal volume.

b **False.** Normal $PaCO_2$–$ETCO_2$ is 5 mmHg (0.7 kPa).

c **True.** The transfer of CO_2 from pulmonary capillaries to alveoli is impaired, e.g. excessive lung inflation (PEEP), COPD. Thus the $PaCO_2$–$ETCO_2$ gradient is increased.

d **True.** Poor pulmonary perfusion results in failure to get rid of CO_2 and relative increase in pulmonary V:Q.

e **True.** The $ETCO_2$ rises in malignant hyperthermia which shows both a respiratory (poor ventilation) and a metabolic acidosis (excess lactate). A similar rise in the $ETCO_2$ occurs in postoperative shivering.

The control of breathing (Qs 29–38)

29 In the central control of breathing:

a Inspiratory neurons are mainly concentrated in the ventral part of the medulla oblongata.

b The apneustic and pneumotoxic centres have opposing effects.

c Inspiratory neurons fire even if isolated from accessory centres.

d The cells of the central respiratory centre are chemoreceptors.

e The expiratory cells are quiescent during normal quiet breathing.

30 In the control of ventilation:

a Breathing increases to the same extent in respiratory and metabolic acidosis, provided the plasma pH falls at the same rate in both.

b The central chemoreceptors respond most of all to H+ ion.

c The composition of the ECF bathing the central chemoreceptors is governed solely by the cerebrospinal fluid (CSF).

d Alterations in $PaCO_2$ take about 60 s to be transmitted to the CSF.

e Hypoxia directly stimulates the central chemoreceptors.

31 In the ventilatory response to CO_2:

a The central chemoreceptors are responsible for most of the response.

b If the $PaCO_2 > 10$ kPa (75 mmHg), then carbon dioxide narcosis occurs.

c The sensitivity of the PCO_2/ventilation response is increased by hypoxia.

d The sensitivity of the PCO_2/ventilation response is decreased in metabolic acidosis.

e The sensitivity of the PCO_2/ventilation response is decreased in sleep.

32 The peripheral chemoreceptors:

a Cells of the carotid and aortic bodies have a very high metabolic rate.

b Blood flow per gram is second only to that of the kidney.

c Carotid and aortic bodies respond to a fall in plasma pH.

d Carotid bodies respond to a rise in local PCO_2.

e contribute to hyperventilation in pernicious anaemia.

33 Concerning peripheral chemoreceptors:

a The carotid and aortic bodies are of about equal importance in man.

b Neural activity from the bodies is abolished if the PaO_2 is raised > 200 mmHg.

c Chemoreceptor activity increases in acute haemorrhage.

d Stimulation of chemoreceptors causes vasodilation in pulmonary blood vessels.

e Stimulation of chemoreceptors causes bronchoconstriction.

34 *The peripheral chemoreceptors:*

a have an insignificant role in the control of normal quiet breathing.
b have an important role in the hyperventilation of moderate exercise.
c have their ventilatory response to hypoxia dampened in acidosis.
d if stimulated, sometimes result in bradycardia.
e have their activity strongly depressed by cyanide.

35 *Concerning the neural control of ventilation:*

a Motorneurons to the diaphragm lack inhibitory Renshaw cells.
b Activity in the swallowing centre inhibits the respiratory centre.
c Ability of an infant to swallow and breath simultaneously is mainly due to neural synchronization.
d Local brain-stem activity is important in stimulating respiratory neurons.
e Different opioid receptors mediate pain and breathing.

36 *The Hering–Breuer inflation reflex (HBIR):*

a is sensitized by steam inhalation.
b prolongs expiratory time.
c receptors are located in the respiratory mucosa.
d receptors are stimulated by the volume but not by the rate of gas inspired.
e is important in regular normal quiet breathing.

37 *Regarding lung reflexes:*

a The Hering–Breuer inflation reflex is sensitized if lung compliance rises.
b The Hering–Breuer deflation reflex is sensed by same receptors as the inflation reflex.
c Head's paradoxical reflex is analogous to the gasp reflex.
d The cough reflex (from larynx, carina, trachea, main bronchi) is only elicited by mechanical stimuli.
e J-receptors are stimulated by both mechanical and chemical stimuli.

38 *The immersion or diving reflex:*

a is characterized by tachycardia.
b is characterized by apnea, laryngeal spasm, and general vasoconstriction.
c may be evoked by dental manipulations.
d develops in childhood.
e may be elicited by splashing one's face with water.

The control of breathing (Answers)

29 a **False.** Inspiratory neurons are mainly concentrated in dorsal regions of the medulla.

b **True.** The apneustic centre tends to prolong inspiration, the pneumotaxic centre tends to switch off inspiration.

c **True.** Inspiratory neurons appear to have the intrinsic property of periodic firing.

d **True.** Although not nearly so sensitive as the nearby distinct central chemoreceptor.

e **True.** Activity in the expiratory neurons only becomes significant during active expiration.

30 a **False.** Ventilation increases more rapidly in respiratory acidosis because CO_2 diffuses so rapidly across into the CSF. Fixed H^+ ion only diffuses slowly into the CSF.

b **True.** The central chemoreceptors are bathed in brain ECF and respond to H^+ of the ECF (which is continuous with the CSF).

c **False.** The central chemoreceptors also respond to local metabolism and local blood flow which also influence local ECF.

d **True.** However, poor CSF buffering means that changes in the local PCO_2 rapidly affect the pH of CSF.

e **False.** Hypoxia directly depresses the central chemoreceptors. Hypoxia may also stimulate indirectly by producing a local acidosis.

31 a **True.** The central chemoreceptors are responsible for up to 80% of the ventilatory response to CO_2.

b **True.** Chemoreceptor response fails completely around a PCO_2 of 13 kPa (100 mmHg). Note that the $PvCO_2$ rises to about 70 mmHg in severe exercise.

c **True.** In this way hypoxaemia increases the central chemoreceptor response to hypercapnia.

d **False.** Sensitivity of PCO_2/ventilation response is increased by a metabolic acidosis and the apnea point (lowest PCO_2 that fails to stimulate breathing) is at a lower $PaCO_2$ than normal.

e **False.** Sleep, narcotics, and general anaesthesia reduce ventilatory sensitivity to CO_2.

32 a **True.** This is despite producing a very low A-V difference.

b **False.** The peripheral chemoreceptors have the highest flow of any body organ; 20 ml/g/min.

c **False.** In humans only the carotid bodies respond to a fall in pH.

d **True.** The carotid bodies may contribute up to 20% of the ventilatory response to increased CO_2.

e **False.** The PaO_2 does not fall in pernicious or in other anaemias.

33 a **False.** The carotid bodies are far more important in man.

b **False.** Some neural activity is present even if PaO_2 is 500 mmHg.

c **True.** Peripheral chemoreceptors are active in haemorrhage because local vasoconstriction causes hypoxaemia and hypercapnia.

d **False.** Activation of peripheral chemoreceptors causes constriction of pulmonary and systemic vessels.

e **True.** Bronchoconstriction induced in this way may be quite important in the fetus.

34 a **True.** Only a minority of chemoreceptor cells are tonically active at normal PaO_2 levels.

b **False.** The arterial blood gases hardly alter, if at all, in moderate exercise.

c **False.** Acidosis and hypercapnia increase the sensitivity of the peripheral chemoreceptors to hypoxia.

d **True.** Bradycardia occurs if the chest is held immobile, e.g. in the fetus. If the chest is free to move one finds tachycardia.

e **False.** Cyanide excites these chemoreceptors, probably by depriving their cells of O_2, unless they too are poisoned by cyanide.

35 a **True.** Thus voluntary effort to the diaphragm is more sustained than those to other muscles.

b **True.** The inhibition of respiration by swallowing helps prevent aspiration of food or fluid.

c **False.** This is mainly due to anatomical differences in the upper airways. Infants are obligate nasal breathers up to 4 months.

d **True.** The respiratory centre is stimulated by the spread of activity from nearby centres, e.g. from the vasomotor centre. Local neural traffic also affects local 'centres'.

e **True.** This gives rise to the possibility of opiate-mediated analgesia that will not depress the respiratory centre.

36 a **False.** Steam appears to blunt the Hering–Breuer inflation reflex.

b **True.** The inflation reflex reduces respiratory frequency.

c **False.** The inflation reflex is mediated via are stretch receptors in the smooth muscle of the large airways.

d **False.** Volume and rate of airflow stimulate these stretch receptors.

e **False.** Bilateral blockade of vagi in adults at rest has no effect on the breathing pattern.

37 a **False.** The inflation reflex is sensitized if pulmonary compliance falls. This may contribute to the rapid shallow breathing of heart failure, croup, and pulmonary congestion. In croup the receptors appear to be directly sensitized.

b **False.** Receptors for the deflation reflex are located deep in the lungs.

c **True.** Sudden inflation of the lungs stimulates further inspiratory efforts.

d **False.** Chemical stimuli are effective in provoking coughing even at very low concentrations.

e **True.** J-receptors will respond to mechanical and chemical stimuli, provided these changes occur in the local environment.

38 a **False.** Vagal slowing of the heart is typical of the diving reflex. Cardiac arrest of up to 20 min has been recorded in the diving seal.

b **True.** Apnea (medullary inhibition); cords approximated (via recurrent laryngeal nerves); vasoconstriction (sympathetic).

c **True.** Apnea, bradycardia, and even cardiac arrest have been recorded occasionally. Also in manipulating endotracheal tubes, ear syringing, bronchoscopy.

d **False.** The diving reflex is better expressed in the fetus (totally immersed).

e **True.** Damping the brow may be more than placebo in treating an emotionally induced supraventricular tachycardia by acting via the cardiac vagus.

Hypoxia, altitude, diving and high pressure (Qs 39–50)

39 *Hypoxaemia:*

a   A PaO_2 of 11 kPa (83 mmHg) is normal in a 60-year-old.

b   Cessation of blood flow to the cerebral cortex may cause permanent damage in only 10–20 s.

c   The heart may be irreversibly damaged if blood supply is stopped for 3–5 min.

d   Short periods of hypoxia may improve cardiac resistance to further hypoxia.

e   Tissue hypoxia can occur in the absence of arterial hypoxaemia.

40 *In hypoxia:*

a   The brain is the most sensitive organ to hypoxaemic damage (PaO_2 < 55 mmHg or O_2 saturation < 88% in a normal subject).

b   low oxygen tensions aggravate any ventilation/perfusion mismatch.

c   Long-term oxygen therapy improves survival in chronic hypoxaemia.

d   Pulse oximetry is most dependable where there is a continuous steady flow of blood in the tissue being monitored.

e   During oxygen therapy, only oxygen flowing early in inspiration reaches the alveoli.

41 *Consequences of hypoxia at altitude (e.g. mountain ascent) include:*

a   tachycardia.

b   respiratory alkalosis.

c   minute volume at rest increases about fivefold at moderate altitude, e.g. 25 000 m (8000 ft).

d   a fall in the PaO_2 to as little as 53 mmHg when the ambient PO_2 is 95 mmHg (altitude of 3500 m or 11 000 ft).

e   noticeable cyanosis.

42 *With hypoxia due to movement to altitude:*

a   The hypoxia directly dilates systemic blood vessels.

b   There is usually a marked rise in arterial blood pressure.

c   A rapid ascent to 2500 m is associated with acute mountain sickness in untrained individuals.

d   Acetazolamide is more useful in treating than in preventing acute mountain sickness.

e   Acute mountain sickness is prevented or reduced by a slow ascent.

43 *In serious altitude-induced illness:*

a   Associated high-altitude cerebral oedema is always preceded by acute mountain sickness.

b   Those with the highest hypoxic ventilatory response are most at risk from cerebral oedema.

c   Acute pulmonary oedema is due partly to massive pulmonary vasodilation.

d   Pulmonary oedema fluid is characteristic of a hydrostatic leak.

e   Up to 10 litres of fluid can accumulate in the tissue spaces of the lung in altitude-induced pulmonary oedema.

44 *During acclimatization to high altitude:*

a   the sensitivity of the central chemoreceptor to CO_2 is blunted.

b   hyperventilation is abolished by breathing oxygen.

c   pH of plasma returns to normal quicker than the pH of CSF.

d   the chronic hypoxaemia increases vascularity in most organs.

e   polycythemia begins within 2 h of exposure to high altitude.

45 *In a breath-held dive to 10 m (33 ft):*

a   the pressure increases by 1 atmosphere for every 10 m (33 ft) one dives beneath the water surface.

b   a relatively normal transfer of oxygen occurs from the alveoli into the blood during the breath-hold dive.

c   there is a relatively normal transfer of CO_2 from blood to alveoli.

d   on average one can breath-hold for about 3 min.

e   loss of consciousness from hypoxaemia can occur if breakpoint postponed (apneic interval prolonged).

46 *About diving:*

a   In snorkel diving intra-alveolar pressure is roughly atmospheric.

b   Intrapleural pressure is more positive than that found at sea level.

c   Snorkel diving is comfortable for up to 30 min at a depth of 5 m.

d   Compressed air should be used at depths greater than 40 m (5 atm).

e   Compressed oxygen should be used at very great depths.

47 *In diving:*

a   the risk of barotrauma is greatest near the surface.

b   pulmonary barotrauma can cause arterial air embolism.

c   middle-ear barotrauma is most common during ascent.

d Eustachian tube patency continues to increase as pressure difference between nasopharynx and middle-ear cleft increases.

e a Valsalva manoeuvre helps open the Eustachian tube.

48 Decompression sickness:

a only occurs during or after ascent.

b Joint pain ('the bends') usually occurs within 1 h of surfacing.

c CNS involvement in decompression sickness is very uncommon.

d Respiratory symptoms ('the chokes') are due to rupture of alveoli.

e Recompression with oxygen and helium is superior to oxygen alone.

49 Using oxygen therapy:

a One may safely breath 100% normobaric oxygen indefinitely.

b A fractional inspired O_2 (FIO_2) of 60–100% in normal man causes generalized vasoconstriction.

c Reactive radicals of oxygen are formed in the body when breathing room air at sea level.

d There are special enzymes in the lung that destroy oxygen radicals.

e One should never administer an $FIO_2 > 24$–28% in patients with chronic respiratory disease.

50 Breathing oxygen at high pressure (OHP):

a can double blood oxygen if breathed at 2 atm.

b is usually delivered between 2 and 3 atm.

c can cause CO_2 narcosis.

d can cause pneumonitis and pulmonary oedema.

e stimulates superoxide dismutase (anti free oxygen radicals).

Hypoxia, high altitude, diving and high pressure (Answers)

39 a **True.** The PaO_2 falls with with ageing. This is probably related to a progressive age-related V:Q mismatch.

b **False.** Cessation of blood flow to the brain may cause permanent global damage after only 20–30 s. Irreversible changes occur in all cases in 3–5 min.

c **True.** It may take the heart up to 10 min to recover after lesser periods of hypoxia.

d **True.** The hypoxic-induced resistance to the harmful effects of subsequent hypoxia is called 'ischaemic preconditioning'.

e **True.** Tissue hypoxia in the presence of a normal PaO_2 can occur in septic shock and in histotoxic hypoxia.

40 a **True.** Visual, cognitive, and electroencephalographic changes develop when oxyHb is less than 89% in normal subjects.

    b **False.** Pulmonary vascular constriction induced by alveolar hypoxia improves V/Q matching.

    c **True.** Up to 40% of patients treated for 1 month with long-term oxygen will find continued supplemental oxygen unnecessary. Oxygen therapy consisting of more than 18 h per 24 has been shown to improve life expectancy in COPD principally by reducing the onset of pulmonary hypertension.

    d **False.** Poor pulse amplitude and poor perfusion are the main causes of failure to obtain a satisfactory signal, apart from poor placement of the sensor.

    e **True.** Oxygen only enters the alveoli during the first sixth of the period of inspiration.

41   a **True.** Tachycardia occurs at altitude as the heart tries to increase oxygen delivery to the tissues.

    b **True.** Hypoxic-driven tachypnoea means CO_2 excretion increases and therefore pH rises.

    c **False.** The resting minute volume is usually only doubled, mainly due to an increase in the depth of breathing. Rate increase only occurs at 10 000–15 000 m.

    d **True.** At 3000 m the alveolar–arterial O_2 difference is only 3 mmHg. None the less, the SaO_2 will be close to 90%.

    e **True.** Detectable cyanosis is present once there is more than 5 g/dl of deoxyHb in the blood from any cause.

42   a **True.** Hypoxic vasodilation may be masked by reflex vasoconstriction (except in heart and brain).

    b **True.** BP increases due to tachycardia despite general systemic hypoxaemic vasodilation. Note that retinal blood flow doubles at 5000 m, hence flame haemorrhages are a common finding.

    c **False.** Acute mountain sickness is not to be expected below 2500 m. The symptoms include headache, nausea, anorexia, dizziness, dyspnoea, and insomnia.

    d **False.** Acetazolamide (a carbonic anhydrase inhibitor; 500 mg/day slow release) is most useful in preventing acute mountain sickness. Dexamethasone (a steroid) is useful in both prophylaxis and treatment.

    e **True.** Modern methods of transportation facilitate a rapid gain of altitude for unacclimatized lowland dwellers, hence many tourists develop this unpleasant but usually short-lived and relatively benign illness. Portable hyperbaric chamber simulates descent and is as useful as drug therapy, probably better (see 42d).

43   a **False.** Preceding symptoms of the benign condition are usual but not invariable.

    b **False.** Those with the highest hypoxic ventilatory drive are the least at risk of altitude-related cerebral oedema because the ensuing hypocapnia causes cerebral vasoconstriction and so inhibits the formation of tissue odema.

c **False.** A fall in PaO_2 and in PAO_2 directly and reflexly cause pulmonary vasoconstriction and this causes pulmonary hypertension.

d **False.** Altitude-induced pulmonary oedema fluid has the characteristics of a permeability leak; that is an exudate (protein content greater than 30 gl) as opposed to a transudate.

e **False.** In pulmonary oedema lung tissue fluid rises to 2–3 l. Normally there is less than 500 ml of tissue fluid in the lungs.

44 a **False.** The sensitivity of the central chemoreceptor is increased. Thus a lower than normal $PaCO_2$ is an effective stimulant.

b **False.** Oxygen does not abolish hyperventilation because the hypoxic drive to breathing via the peripheral chemoreceptors is no longer paramount in acclimatized people.

c **False.** Plasma pH is normalized by renal excretion of HCO_3^- (slow); CSF pH is normalized probably by secretion of HCO_3^- into plasma (fast).

d **True.** Vascularity increases in organs and there is also an increase in the number of mitochondria in cells.

e **True.** Over the succeeding 4–6 weeks of living at a high altitude, Hb may rise to 20 g/dl.

45 a **True.** Pressure does not increase to the same extent in tunnels and mines as it does under water. This has importance in diving as air usage increases with depth. Thus at 10 m a supply of air will last half as long as just below the surface.

b **True.** Alveolar pressure (PAO_2) rises to 24 kPa (180 mmHg) in free diving.

c **False.** Reversed transfer of CO_2 occurs in free diving. The $PACO_2$ rises to 6 kPa ($PaCO_2$ is 5.3 kPa).

d **False.** Normal breath-hold is about 1.5 min. The rising CO_2 is the major breakpoint determinant.

e **True.** In preparatory hyperventilation, hypoxaemic syncope can occur before the $PaCO_2$ has time to build up to a breakpoint.

46 a **True.** Alveoli are in open connection with air above the water when snorkelling.

b **False.** IP pressure is far more negative than normal. Otherwise lungs cannot remain expanded in face of increased external pressure. Unlike in humans, bronchial cartilage extends to the alveoli in seals, thus preventing alveolar collapse.

c **False.** Maximum IP pressure one can generate is about −100 mmHg. Snorkelling for any reasonable time is only comfortable at a depth of about 0.5 m.

d **False.** Beyond 40 m in depth significant absorption of N_2 occurs, leading to narcosis. N_2 is replaced by helium (heliox) alone or with hydrogen (trimix).

e   **False.** One should reduce the O_2 concentration for progressively deep dives. O_2 is neurotoxic at 4 atm. The danger increases with the duration of exposure and if the subject exercises. Increasing depth means increased absorption of O_2. Pure oxygen must never be used in diving cylinders.

47   a   **True.** Near the surface a small change in depth causes a relatively large change in volume.

b   **True.** Pulmonary barotrauma occurs during ascent. It is the most serious form of barotrauma.

c   **False.** Middle-ear barotrauma is most common in descent. It is the most common barotrauma in divers.

d   **False.** If pressure difference exceeds 90 mmHg the Eustachian tube cannot open.

e   **False.** A Valsalva manoeuvre results in further locking forces. May lead to rupture of the tympanic membrane or to inner-ear barotrauma.

48   a   **True.** Decompression sickness is due to liberation of N_2 from super-saturated tissues at a lower absolute ambient pressure.

b   **True.** The elbow or shoulder is involved three times more often than the knee or hip.

c   **False.** CNS involvement is common even in mild cases.

d   **False.** Gas bubbles in pulmonary arterioles; even 5 ml causes retrosternal discomfort ('the chokes').

e   **True.** N_2 is more soluble than helium in structural fats of the CNS. Recompression with O_2–helium allows faster elimination of N_2.

49   a   **False.** Inhalation of 100% O_2 at 1 atm causes substernal pain and damage to the alveolar/capillary membrane within 24 h.

b   **True.** 60–100% O_2 causes vasoconstriction, especially in cerebral and retinal vessels. Optical phenomena and light-headedness often occur.

c   **True.** H_2O_2, superoxide, O_2 singlet are all formed at sea level.

d   **True.** Lungs are rich in catalase, peroxidase, superoxide dismutase.

e   **False.** 100% O_2 is often necessary in acute respiratory illness but care must be taken that it is not administered to patients with COPD who retain CO_2 and rely on hypoxic drive.

50   a   **False.** OHP will not increase amount of oxygen carried by Hb, as at 1 atm it is on the plateau of its saturation curve already. Normally 0.3 ml O_2 is dissolved in blood; > 4.0 ml becomes dissolved at 2 atm OHP. For each mmHg of PO_2 there is 0.003 ml O_2 dissolved in blood.

b   **True.** At 3 atm OHP 6 ml of oxygen is dissolved in 100 ml blood.

c   **True.** Since there is little deoxygenated Hb at an OHP of 3 atm CO_2 cannot be buffered properly and thus it builds up in the tissues.

d   **True.** However, the greatest danger in OHP is fire from an electrical spark.

e   **True.** This effect of OHP is important in crush injuries, grafts and replants where free radical formation causes reperfusion injury. Note that neither

OHP nor other attempts to provide oxygen delivery values greater than 650 ml/min/m² ('supranormal haemodynamics') reduces morbidity or mortality in critically ill patients.

Pulmonary circulation (Qs 51–54)

51 *In the pulmonary and bronchial circulation:*

    a At any instant the lungs contain about 250 ml of blood (subject at rest and upright).

    b When erect there is a very poor flow in bronchial vessels at the apex of the lung.

    c About 2% of mixed venous blood is normally shunted to the systemic side.

    d The bronchial system does not anastomose with the pulmonary one.

    e The bronchial circulation is mainly nutritive to lung tissue.

52 *Within the pulmonary circulation:*

    a The pulmonary capillary pressure is normally about 6–7 mmHg.

    b There is about 100 ml of blood in the pulmonary capillaries (at rest, upright).

    c Pressure in pulmonary capillary must exceed that in alveolus for blood flow to occur.

    d Flow ceases in apical vascular capillaries even if intra-alveolar pressure only rises by 5 mmHg.

    e There is little or no tissue fluid in normal lungs.

53 *In the pulmonary circulation:*

    a Pulmonary hypertension is defined as a mean pulmonary artery pressure of > 20 mmHg.

    b Pulmonary artery wedge pressure (PAW) is the same as the left atrial pressure (LAP).

    c The plasma colloid osmotic pressure (COP) minus PAW should exceed 8 mmHg in normal people.

    d It is possible to have a pressure of 90 mmHg in pulmonary artery and 9 mmHg in pulmonary capillaries at the same time.

    e Tissue fluid tends to track along the peribronchial spaces.

54 *With pulmonary hypertension:*

    a Pulmonary oedema is the most important consequence of pulmonary hypertension.

    b Right ventricular hypertrophy occurs in chronic cases even if the cause of pulmonary hypertension is in the left side of the heart.

    c Acute right ventricular hypertrophy occurs if there is a sudden rise in pulmonary artery pressure.

d Endothelial dysfunction is prominent in primary cases of pulmonary hypertension.

e Endothelial dysfunction is prominent in secondary cases of pulmonary hypertension.

Pulmonary circulation (Answers)

51 a **False.** The lungs contain about 450–500 ml blood in the upright position.

b **False.** Bronchial vessels have a high driving pressure (arise from the aorta).

c **True.** About 2–4% is shunted. Thus arterial blood can never be 100% saturated with oxygen.

d **False.** The pulmonary and bronchial systems communicate via numerous anastomoses at precapillary level.

e **True.** In this way the bronchial circulation is comparable to the hepatic circulation in the liver.

52 a **True.** Pulmonary capillary pressure is roughly halfway between the mean pulmonary artery pressure (13 mmHg) and the left atrial pressure (1–2 mmHg).

b **False.** Pulmonary capillaries normally contain about 50 ml of blood.

c **True.** Flow ceases in high PEEP, at very high tidal volumes, and during the Valsalva manoeuvre.

d **True.** Normally apical pulmonary capillary flow is negligible (erect).

e **False.** There is about 500 ml. Note that the subatmospheric (negative) pressure in lung tissue spaces tends to draw fluid out of the capillaries.

53 a **True.** Alternative definition of pulmonary hypertension: a pulmonary artery systolic pressure greater than 30 mmHg.

b **False.** PAW is normally about 5–7 mmHg, slightly more than the LAP although clinically it is a good estimate of LA filling pressures.

c **True.** If COP–PAW < 8 mmHg then pulmonary oedema results.

d **True.** If the arterioles remain constricted then very high pressures may be found in the pulmonary artery, whereas the pulmonary capillary pressure may not be inordinately elevated. (Note morbid obesity.)

e **True.** Fluid also tracks along the perivascular spaces. Excess causes interstitial oedema.

54 a **True.** Pulmonary oedema is life threatening. Note that right ventricular dysfuction, hepatic failure, and pulmonary infarction may also occur.

b **True.** Pressure rise in the left side of the heart is eventually transmitted back to the right side, e.g. in chronic mitral stenosis.

c **False.** Right ventricular hypertrophy only occurs in long-standing cases.

d **True.** In primary cases there is a notable a fall in endothelial nitric oxide and PGI_2 with a rise in endothelin.

e **True.** Endothelial dysfunction is a result, rather than a cause, of secondary pulmonary hypertension.

Drowning, sleep apnoea and COPD (Qs 55–59)

55 Drowning:

a More drownings occur in fresh water than at sea.

b Almost half the cases die from laryngeal spasm and subsequent hypoxia.

c In freshwater aspiration there is massive pulmonary oedema.

d Freshwater aspiration causes massive intrapulmonary shunting.

e Up to 25% of those resuscitated after near-drowning die within the succeeding 24 h.

56 The following statements are true about sleep apnoea:

a A normal night's sleep is interrupted by up to 10 apnoeas, each lasting less than 10 s.

b In central sleep apnoea there is an absence of respiratory efforts.

c There may be up to 500 apnoeas/night in obstructive sleep apnoea.

d Obstruction is between the larynx and carina in obstructive apnoea.

e Sex distribution is predominantly male in all types of sleep apnoea.

57 Sleep apnoea:

a is frequently complicated by pulmonary hypertension.

b is unusual in that secondary polycythemia is rarely a sequel.

c is commonly accompanied by narcolepsy.

d is improved, and the frequency of attacks may be lessened, by taking moderate amounts of alcohol at night.

e may be improved, and one may help prevent apnoeic episodes, by oxygen therapy at night.

58 The following are true about factors in pulmonary disease:

a The incidence of COPD is inversely proportionaly to the intake of fruit and vegetables.

b There is good evidence to show that a high fish-oil intake may protect against COPD.

c A high sodium intake has been proved to worsen asthma, especially in men.

d Exercise in COPD is limited by the maximal rate of O_2 transport to the mitochondria of exercising muscles.

e Arm exercises are more fatiguing than similar-level leg exercises in those with chronic lung disease.

59 The following miscellaneous features about the lungs are true:

a Air becomes supersaturated with H_2O vapour in the respiratory passages.

b The lungs contribute little to autotransfusion in acute blood loss.

c Alveoli are a good source of angiotensin-converting enzyme (ACE).

d   Inositol supplementation is useful in perinatal respiratory distress.

e   Alpha 1 Trypsin inhibitor is formed in the pulmonary endothelium.

Drowning, sleep apnoea, and COPD (Answers)

55   a   **True.** About 75% of drownings occur in fresh water, 40% in under 5-year-olds.

b   **False.** Dry asphyxia occurs in only 10–20% of cases of drowning.

c   **False.** In freshwater aspiration there is massive haemolysis. The blood volume increases 50% in about three minutes. Saltwater aspiration causes pulmonary oedema.

d   **True.** Freshwater aspiration also damages surfactant, causes alveoli to collapse, and causes a fall in the V:Q.

e   **True.** Deaths after near-drowning are usually from diffusion block, metabolic acidosis, and hypothermia.

56   a   **True.** Normal pattern of sleep apnoea is sometimes called physiological Cheyne–Stokes breathing.

b   **True.** There is no respiratory effort despite a rise in $PaCO_2$ and a fall in the PaO_2.

c   **True.** The subject does not remember these apnoeas despite their frequency. However, they suffer daytime fatigue and are prone to dozing.

d   **False.** The obstruction in obstructive apnoea is in upper airways between the nose and larynx, i.e. the pharynx.

e   **False.** Central sleep apnoea has about an equal sex distribution. However, obstructive sleep apnea is predominantly in obese males.

57   a   **True.** Pulmonary hypertension develops because of the repeated hypoxic episodes characteristic of sleep apnoea.

b   **False.** Secondary polycythemia is an important and frequent complication.

c   **False.** Narcolepsy is a disorder of REM sleep patterns.

d   **False.** Alcohol also makes efforts at arousal more difficult.

e   **False.** Oxygen therapy lessens ill-effects of apnea. None the less, remember that hypoxaemia is a powerful stimulus to breathing. PEEP masks are useful in obstructive apnea as they help to keep the airways open.

58   a   **True.** Most studies have shown that fruit is protective against COPD. The reason is not known. It may be the antioxidant effect of vitamin C.

b   **True.** Fish-oils have antioxidant effects. Unopposed oxidant activity is important in the pathogenesis of lung damage in COPD.

c   **False.** It is currently thought that the 'sodium effect' in asthma may be due to a coincidental decreased Mg intake. Mg relaxes airways. However, there is some conflicting evidence.

d   **False.** Exercise in COPD is limited by the maximal level of ventilation the patient can sustain.

e **True.** Shoulder girdle involvement restricts respiratory movements, especially in those with large functional residual capacity (little diaphragmatic contribution to inflation).

59 a **False.** Air in the lung is saturated with water vapour at local lung temperature.
b **False.** The lungs have a mobilizable reservoir of 200–300 ml of blood.
c **False.** ACE is found mainly in the pulmonary vascular endothelium.
d **True.** Inositol helps control surfactant phospholipid synthesis.
e **False.** Alpha 1 Trypsin inhibitor is formed solely in the liver. It inhibits elastase in the lungs.

7

Gastrointestinal

Structure and innervation (Qs 1–5)

1 The following are true of the gastrintestinal tract (GIT):

a All the muscle in the GIT is smooth muscle.

b After death the length of the intestine is shorter than *in vivo*.

c The small intestine can respond to stretch by an increase or a decrease in tension.

d Elimination of waste products from the body is a major function of the GIT.

e The muscularis mucosa is composed of an outer longitudinal and an inner circular layer.

2 Regarding the structure of the GIT:

a The small intestine is lined by a single-layered columnar epithelium.

b The entire lining of the small intestine is replaced in 5–6 days.

c Small intestine mucosa is replaced by cells from the crypts of Lieberkühn.

d Crypt cells are mature enterocytes.

e The apex of a small intestine villus is hypoxic.

3 In terms of the parasympathetic supply of the GIT:

a cholinergic activity generally increases activity in the smooth muscle of the GIT.

b nerves end in the myenteric (Auerbach's) plexus and in the submucous plexus (Meissner's plexus).

c Afferent fibres that travel with the parasympathetic end in a similar way to general somatic afferent nerves.

d Acetylcholine (ACh) is the major secretion of the vagus in the GIT.

e The vagus supplies the GIT from the lower oesophagus to the hepatic flexure.

4 The sympathetic supply of the GIT:

a is generally secretomotor in the intestine.

b causes powerful vasoconstriction throughout the intestinal bed.

c relaxes the muscularis mucosa.

d releases noradrenaline (NA) only at the termination of its postganglionic fibres in the GIT.

e blocks the release of mediators from parasympathetic terminals in the myenteric plexus.

5 Concerning receptors and transmitters in the GIT:

a Mechanoreceptors are found in mucosa, submucosa, and muscle layers.

b Osmoreceptors are present from the duodenum to the proximal ileum.

c Circulating ACh has significant effects.

d   Endocrine–paracrine cells are scattered in the submucosa of the small intestine.

e   Endocrine–paracrine cells store and secrete substances that are also found in neural endings of the enteric nervous system.

Structure and innervation (Answers)

1   a   **False.** Voluntary muscle is present in oesophagus and rectum.

b   **False.** The GIT is longer after death because some contractual tone is always present during life.

c   **True.** Relaxation or contraction occurs depending on the initial length of the fibres and the nature and bulk of the distending material.

d   **False.** The kidney is the major eliminator of body waste. The lung and liver also excrete waste.

e   **True.** A similar arrangement is found in the muscularis mucosa as in the main muscle layers of the GIT.

2   a   **True.** The intestinal mucosa is infolded to form villi. Ideal for absorption and secretion.

b   **True.** GIT mucosal cells are among the most rapidly replicating in the body.

c   **True.** Over 2–5 days crypt cells migrate up the sides of the villi.

d   **False.** Digestive enzymes and transport proteins are inserted into the cell membrane during migration up the sides of the villi as part of continuing maturation.

e   **True.** The apex of the villus is anoxic due to the countercurrent flow of O_2 from artery to vein within the villus.

3   a   **True.** The parasympathetic system also affects secretomotor neurons of the intrinsic enteric nervous system, causing the release of relaxatory polypeptides, e.g. VIP, somatostatin, and encephalins, and excitatory ones, e.g. substance P, galanin.

b   **True.** Parasympathetic terminals end in both Auerbach's and Meissner's plexus and all are preganglionic in type.

c   **True.** Strictly speaking, the autonomic system is a purely efferent system.

d   **True.** Other transmitters and cotransmitters include dopamine, purines, 5-HT, peptides (substance P, VIP).

e   **True.** The vagus supplies the GIT from the upper oesophagus to mid transverse colon, i.e. the lower oesophagus and the hepatic flexure are included within this area.

4   a   **False.** The sympathetic inhibits gut contraction and constricts some sphincters.

b   **True.** The splanchnic vascular bed is extremely responsive to neural activity.

c **False.** The sympathetic contracts the muscularis mucosa and excites local mucus glands.

d **False.** NA is released along the course of the terminal nerve fibre.

e **True.** The blocking of ACh release is in addition to the direct effect of NA released locally.

5 a **True.** Mechanoreceptors are are especially involved in the co-ordination of movements of the GIT.

b **False.** Gut osmoreceptors are found only in the mucosa of the duodenum and upper jejunum.

c **False.** Plasma ACh is rapidly degraded by plasma cholinesterases.

d **False.** Endocrine–paracrine cells are present in the mucosa beside ordinary enterocytes.

e **True.** The secretory granules in both endocrine–paracrine cells and in cells of the enteric nervous system contain many compounds common to both.

Movements (Qs 6–16)

6 *Swallowing:*

a The oral stage is the only voluntary stage of swallowing.

b Swallowing is only activated by stimulation of receptors in the mouth or oropharynx.

c Vagal innervation is essential for the orderly propagation of the primary peristaltic wave along the oesophagus.

d The resting pressure in the cricopharyngeal sphincter exceeds that in the lower oesophageal sphincter (LOS).

e The oesophageal stage is the quickest stage of swallowing.

7 *Concerning cardiac competence and the lower oesophageal sphincter (LOS):*

a The pressure in the lower 2 cm of the oesophagus rises as the intra-abdominal pressure rises.

b The LOS is tonically contracted when one is not swallowing.

c Vasoactive intestinal polypeptide (VIP) relaxes the LOS.

d Mechanical factors are of critical importance in maintaining cardiac competence of the LOS.

e Achalasia and megaoesophagus may be associated with damage to the vagi.

8 *Vomiting:*

a Vomiting is usually associated with retrograde peristalsis.

b The vomiting centre is situated in the hypothalamus.

c Emetics directly activate the vomiting centre.

d Ipecacuanha mainly affects receptors outside the CNS.

e Concomitant closure of the glottis is an integral part of vomiting.

9 Concerning nausea, vomiting, and retching:

a A separate nausea centre exists in the medulla.

b Salivation almost always accompanies vomiting.

c In retching there is an annular constriction between the body and antrum of the stomach.

d In retching there is a simultaneous relaxation of the upper and lower oesophageal sphincters.

e Mucosal tears are likely in repeated retching.

10 About stomach motility:

a There is a higher frequency basal electrical rhythm (BER) in the fundus than in the antrum.

b Motilin decreases the basal electrical rhythm.

c Receptive relaxation only occurs in the proximal stomach.

d The pylorus is tonically contracted when the stomach is empty.

e Strong antral contractions may result in contraction of the pylorus.

11 Gastric emptying:

a is faster after small-volume meals than after large ones.

b is increased by the products of protein digestion.

c is reduced by the products of fat digestion.

d is inhibited by gastrin.

e can be either increased or decreased by osmoreceptors in the duodenal mucosa.

12 Segmentation movements in the small intestine

a are the most important movements for the mixing of food.

b cause some aboral propulsion of food.

c are more frequent in the duodenum than in the ileum.

d the length of individual segments in segmentation increases progressively from the duodenum onwards.

e can occur in completely denervated gut.

13 Peristalsis in the small intestine:

a occurs in denervated intestine.

b is abolished by painting the mucosa with a local anaesthetic.

c takes the form of strong peristaltic rushes from time to time.

d can be initiated by a myoelectric complex in the antrum/duodenum.

e and in particular long-distance peristaltic reflexes, are subject to the control of extrinsic nerves.

14 With movements of the large intestine:

a Peristaltic movements have a similar basis to those elsewhere.

b Mass propulsion can be independent of extrinsic nerves.

c Haustral movements are as likely to cause retrograde as orthorograde progression.

d The rectum is normally firmly contracted.

e The resting rectal pressure tends to drive material into the sigmoid colon.

15 *In defecation:*

a the desire to defecate is related to intrarectal pressure.

b rectal distension is signalled by mechanoreceptors in the mucosa of the anorectal region.

c rectal distension signals are carried by afferents to the lower lumbar cord.

d the smaller the stool the less the pressure needed to expel it.

e involuntary defecation depends on an intact reflex arc.

16 *Gut motility:*

a The normal recovery time of a marker from mouth to anus is about 72 h.

b Constipation implies the infrequent passage of stools.

c A subdiaphragmatic truncal vagotomy is associated more with diarrhea than with constipation.

d The dumping syndrome is likely to follow a truncal vagotomy unless a pyloroplasty or gastrojejunostomy is also done.

e Aganglionosis of colon (Hirschsprung's disease) causes megacolon.

Movements (Answers)

6 a **True.** The pharyngeal and oesophageal stages are involuntary.

b **True.** One must voluntarily or involuntarily stimulate receptors in these areas in order to initiate the swallowing reflex. Swallow all the saliva in your mouth (3–4 swallows) and it becomes impossible to start another swallow.

c **True.** Orderly effective propagation of the primary peristaltic wave in the oesophagus will not occur if the vagi are damaged.

d **False.** The resting pressure in the upper sphincter is 50–100 mmHg, in the LOS it is 15–40 mmHg.

e **False.** In swallowing the first two stages only last < 2 s whereas the oesophageal stage lasts 8–10 s.

7 a **True.** Normally the lower 2 cm of the oesophagus is below the diaphragm and is subject to changes in intra abdominal pressure.

b **True.** None the less the LOS is a functional not an anatomical sphincter.

c **True.** The VIP comes from special vagal fibres, or is co-released by the vagus, or comes indirectly via the enteric plexus or via associated non-adrenergic non-cholinergic nerves (NANC). NO also relaxes the LOS. VIP/NO relaxation is balanced by the constrictor effects of ACh and substance P.

d **True.** Cardiac competence is maintained by the acute angle of entry of the oesophagus into the stomach, occlusive valve-like folds of the mucous membrane, tension on the phreno-oesophageal ligaments, and a positive intra-abdominal pressure.

    e **True.** In Chagas disease trypanosomes attack the postganglionic fibres in the oesophageal wall. Note the occasional use of intrasphincteric botulinum toxin (prevents release of ACh) in achalasia.

8   a **False.** Usually the stomach and oesophagus are flaccid in vomiting. The force to expel vomit is provided by abdominal muscles.

    b **False.** The vomiting centre is in the medulla ventral to the nucleus of the solitary tract.

    c **False.** Many emetics act on the area postrema, outside the blood–brain barrier in the floor of the IVth ventricle.

    d **True.** Ipecacuanha causes vomiting by stimulating duodenal receptors.

    e **True.** This protects against aspiration of vomitus (acid aspirate particularly harmful to lung, Mendelson's syndrome). Note that the epiglottis is not essential for normal swallowing.

9   a **True.** Non-nauseous vomiting can occur, e.g. psychogenic, cerebral.

    b **True.** Salivation helps to lubricate material vomited and dilutes acid, minimizing dental damage.

    c **True.** Antrum contracts, thus driving material into the limp fundus and lower oesophagus.

    d **False.** The cricopharyngeus is tightly constricted in retching.

    e **True.** Deep inspiration of a retch creates excessive negativity around herniated parts of stomach and may tear the mucosa (Mallory–Weiss tear).

10   a **False.** BER from fundus to antrum is 3–4/min; BER from antrum to pylorus is 15–20/min.

    b **False.** Motilin (from enterochromaffin cells of gut) increases the rate of the BER.

    c **True.** Relaxation of the proximal stomach is due to special properties of the oblique muscle in this area.

    d **False.** The pylorus is relaxed but closed. Pyrloric relaxation is due to local nitric oxide. A deficiency of NO synthase occurs in congenital pyloric stenosis.

    e **True.** Antral contractions result in retropulsion, churning, and mixing of food. Note the small lumen of the pylorus greatly facilitates retropulsion.

11   a **False.** Large-volume meals engender large tension and thus faster emptying than with small-volume meals.

    b **False.** Products of protein digestion slow the emptying rate.

    c **True.** Products of fat digestion slow emptying via the liberation of cholecystokinin (CCK) from the intestinal mucosa. CCK inhibits the amplitude of gastric contractions.

    d **False.** Gastrin stimulates antral contraction directly and via local cholinergic plexuses. This causes chyme to squirt into the duodenum in discrete bursts.

    e **True.** Hypertonic chyme in the duodenum slows gastric emptying; hypotonic chyme speeds gastric emptying.

12 a **True.** Localized annular constrictions divide and subdivide discrete high-pressure areas.

   b **True.** Some aboral propulsion takes place because of different frequency of segmentation in different areas.

   c **True.** Segmentation frequency is 10–12/min in duodenum and 6–8/min in ileum.

   d **False.** Segment lengths are 8–10 cm in duodenum and 0.5–1 cm in ileum.

   e **True.** Note that segmentation is enhanced by parasympathetic and inhibited by sympathetic nerves.

13 a **True.** Peristaltic movements can occur in denervated gut but extrinsic nerves and hormones can initiate and modify them.

   b **False.** Non-mucosal distension receptors can initiate peristalsis.

   c **False.** A peristaltic rush occurs only in the large intestine.

   d **True.** When fasting, the myoelectric complex initiates a pattern of movement (segmentation, peristalsis) and of quiescence.

   e **True.** Distension of one part of the gut may inhibit activity in another (intestino-intestinal inhibitory reflex). Food in the stomach may stimulate activity in the small intestine (gastro-intestinal reflex).

14 a **True.** Peristalsis is less freqeuent in the large gut than in the small gut.

   b **True.** Hormones produced elsewhere affect large gut movement, e.g. CCK may help in the gastrocolic reflex (which is thus both neural and hormonal in nature).

   c **True.** Haustrations are due to annular constriction of circular muscle. They last 50 s and generate 50 mmHg of pressure. Their genesis not understood.

   d **False.** The rectum is normally empty and its walls are collapsed.

   e **True.** The resting rectal pressure is the basis for the retrograde passage of suppositories.

15 a **True.** The desire to defecate becomes apparent at an intrarectal pressure of 20 mmHg; it becomes intense at 50 mmHg.

   b **True.** A normal anorectal mucosal junction is critical for the normal defecation reflex.

   c **False.** Rectal distension signals are carried to upper sacral cord (mainly S2).

   d **False.** Bulk foods are recommended for patients in whom straining at stool may cause problems, e.g. piles, hypertension.

   e **True.** Involuntary defecation is a reflex. Also one needs intact local control. This occurs after cord transection. Control of the reflex is learnt as an infant in 'toilet training'.

16 a **True.** Transit times to: caecum, 4 h; hepatic/splenic flexures, 6/8 h; sigmoid colon, 12 h; expulsion, 72 h.

   b **False.** Constipation implies the difficult passage of hard stools and/or a stool frequency noticeably less than usual for a particular person.

   c **True.** A truncal vagotomy causes a loss of exocrine pancreatic secretion, thus there is fat malabsorption and this causes steatorrhea.

   d **False.** The dumping syndrome is typical of postgastrectomy. It is probably due to the rapid entry of hyperosmotic fluid into the small intestine, consequent ECF depletion, and possibly the entry of hypotensive kinins from gut to blood.

   e **True.** Aganglionic segment(s) will not transmit propulsive movements.

Secretions (Qs 17–33)

17 *Saliva:*

   a Normally the secretion of saliva is less than 600 ml/day.

   b Saliva secreted by the acini is about isotonic with plasma.

   c The salivary ducts are relatively impermeable to water.

   d The pH of resting saliva is higher than that of stimulated saliva.

   e K^+ level in copious stimulated saliva is less than plasma K^+.

18 *Regarding saliva:*

   a CO_2 escapes from saliva once it enters the mouth.

   b Salivary secretion is controlled by nerves and hormones.

   c Psychic input can have a very significant effect on salivary flow.

   d Parasympathetically induced vasodilation in salivary glands is blocked by atropine.

   e Parasympathetically induced secretion of saliva is blocked by atropine.

19 *During salivation:*

   a Sympathetic activity opposes the parasympathetic.

   b Acinar secretory cells actively secrete chloride into primary salivary ducts.

   c The parotid glands normally provide most of the salivary volume.

   d Salivary amylase is added mainly by the duct cells.

   e Plasma iodide is trapped by the salivary glands.

20 *Regarding dental caries:*

   a The high buffering capacity of saliva helps prevent caries.

   b Salivary Ca^{2+} and PO_4^- promote remineralization of lytic dental lesions.

   c The surface of a tooth is normally covered with a layer of complex salivary proteins.

   d The critical pH for the deposition of dental plaque is pH 5.5.

   e The anticariogenic effect of saliva is maximal in resting secretions.

21 *Functional anatomy of the stomach:*

   a Parietal cells are found in the cardia, fundus, and body.

   b Parietal cells are located mainly in the bases of the gastric glands.

   c Mucous cells secrete both mucus and bicarbonate.

d   The pH immediately adjacent to surface epithelium is neutral even during acid secretion.

e   The 'empty stomach' contains up to 100 ml of material.

22 In gastric secretion:

a   the cephalic stage of secretion is responsible for up to half of the acid response to a normal meal and lasts about 1 hour.

b   the cephalic stage in man is stimulated by the thought or sight of food.

c   vagal activity causes the release of gastrin.

d   the products of digestion have little effect on gastric secretion.

e   distension of the stomach inhibits further gastric secretion.

23 In gastric secretions:

a   Pure parietal juice is equivalent to 0.5 M HCl.

b   Non-parietal juice is rich in NaCl and HCO_3^-.

c   Potassium is present only in parietal juice.

d   Basal acid secretion is about 10% of full active secretion.

e   The Zollinger–Ellison syndrome is usually due to gastrinoma in the antrum.

24 About parietal cell function:

a   Parietal cells are packed with tubulovesicular and canalicular structures.

b   Histamine is released by the terminals of local nerves in the stomach.

c   Histamine increases cAMP production in the parietal cell.

d   Prostaglandins synergize with histamine in the parietal cell.

e   Gastrin and ACh stimulate cells mainly via calcium influx.

25 In the generation of gastric acid:

a   The 'proton pump' (H^+,K^+-ATPase) is on the basolateral walls of the parietal cell.

b H^+ is made available for the pump by exchange for Na^+ on the abluminal wall of the parietal cell.

c Cl^- is transported by a primary chloride pump.

d K^+ for the proton pump is K^+ that has left via a conductive path to the lumen.

e   Acetazolamide (carbonic anhydrase (CA) inhibitor) dramatically reduces acid secretion.

26 In the management of peptic ulcer:

a   antimuscarinic agents are ineffective in reducing acid secretion.

b   somatostatin is a potent inhibitor of acid secretion.

c H_1 receptor antagonists can reduce gastrin-mediated acid secretion.

d   when anacidity in the canaliculus is achieved, substituted benzimidazoles, e.g.omeprazole, cease to accumulate in the parietal cell.

e   Omeprazole diminishes acid secretion by any secretagogue.

27 Regarding peptic ulcer:

a   The maxim 'no acid no ulcer' is no longer valid in light of current ideas.

b   It has now been shown that the size and activity of the secretory cell mass is increased in virtually all peptic ulcer patients.

c   Antacid drugs and drugs that produce anacidity can cause hyper-gastrinaemia.

d   *H. pylori* infection is nowadays widely accepted as the most common known cause of peptic ulcer.

e   Detection of labelled exhaled CO_2 after ingesting urea indicates H. pylori infection.

28 In pancreatic exocrine function:

a   the secretory flow rate from acini is relatively small compared to that from the ducts.

b   secretin provokes a very alkaline juice (pH about 8.0).

c   most of the HCO_3- secreted comes from the acini.

d   acini produce more protein per gram than any other tissue.

e   ducts produce a trypsin inhibitor.

29 About pancreatic exocrine function:

a   The intestinal phase is much more important than the cephalic and/or gastric phase in stimulating the exocrine pancreas.

b   A low pH in the duodenum results in a major acinar secretion.

c   Intestinal fat products and amino acids are major secretagogues for proenzyme secretion.

d   Patients with the Z–E syndrome have malabsorption of fat and a relative pancreatic lipase deficiency.

e   Hypocalcaemia is often found in acute pancreatitis.

30 Bile secretion:

a   The liver secretes about 500–1500 ml of bile per day.

b   Only one-third of liver bile comes from the hepatocytes.

c   The level of bile salts in portal blood is the chief controller of duct and of hepatocyte biliary secretion.

d   Non bile acid dependent bile is chiefly controlled by hormones.

e   The liver forms about 20 g of cholic and chenodeoxycholic acid per day.

31 Bile:

a   Primary bile acids are conjugated in bile ducts to form bile salts.

b   Most of the inorganic constituents of bile come from the duct system.

c   Bile is concentrated five- to twentyfold in the gall bladder.

d   Osmotic pressure increases considerably in the gall bladder due to the large-scale concentration of its contents.

e   Gall bladder bile is more acid than hepatic bile.

32 *In the turnover of bile:*

a all micelles are arranged so that the fat-soluble poles are dissolved in the core of the micelle.

b water is reabsorbed by a primary active process from the gall bladder.

c most of the secreted bile salts are reabsorbed in the terminal ileum.

d Bile salts do not normally appear in the faeces.

e More bile salt enters the intestine per digestive period than is present in the total hepatic bile salt pool.

33 *Gall stones:*

a Infection is a prerequisite for the formation of all gallstones.

b Phospholipids increase the solubility of cholesterol.

c The normal cholesterol:bile salts ratio should be < 1:20.

d Calcium carbonate stones may occur if there is an acidification defect in the gall bladder.

e Gall stones are most likely to be formed at night or if meals are skipped.

Secretions (Answers)

17 a **False.** Normally 1.0–1.5 litres of saliva are secreted per day.

b **True.** Primary acinar secretion is subject to electrolyte reabsorption in ducts (NaCl in exchange for $KHCO_3$).

c **True.** Saliva becomes hypotonic in the duct system.

d **True.** The pH of resting saliva is higher than that of stimulated saliva because of the secretion of $KHCO_3$ by the ducts. None the less, when mixed with mouth contents (bacteria, cells, food remnants) the pH in the resting mouth is usually in the range 6.35 to 6.8.

e **False.** The K^+ concentration is 30 mmol/l if the rate of secretion is 0.5 ml/min; K^+ = 18 mmol/l at 4 ml/min. Na^+ in resting saliva is only 5 mmol/l.

18 a **True.** $KHCO_3 \Leftrightarrow CO_2 + KOH$. This allows the mouth pH to rise.

b **False.** The secretion of saliva is controlled by autonomic nerves.

c **True.** The thought or sight of food has little, if any, positive effect on salivation *in man*, but emotions, e.g. fear, can cause a marked inhibition of salivation.

d **False.** Vasodilation in the salivary glands may be due to bradykinin, VIP, local hyperosmolarity, purine release, or a combination of these factors.

e **True.** Salivary secretion is due to activation of muscarinic receptors by ACh.

19 a **False.** The sympathetic contracts myoepithelial cells of the salivary duct. It also increases intraductal K^+ and HCO_3^-.

b **True.** The frusemide-sensitive $Na^+/K^+/2Cl^-$ channels on the basolateral walls of acinar cells facilitate the passage of Cl^- into the lumen via luminal wall chloride channels.

c **False.** Submandibulars provide 70% of salivary volume, parotids 20%, and sublinguals 5%.

d **False.** Duct cells add lysozyme, growth factors, nucleases.

e **True.** Plasma:saliva ratio of iodide may reach 1:60. Significance is uncertain.

20 a **True.** $NaHCO_3$, $KHCO_3$ and salivary proteins help neutralize acids produced on teeth surfaces.

b **True.** Saliva is supersaturated with calcium and phosphate.

c **True.** Bacteria adhere to this layer around teeth.

d **False.** When the pH of plaque is less than 5.5 demineralization begins. Thus pH 5.5 is called the 'critical pH'.

e **False.** Anticariogenic effects are maximal at high rates of flow which wash out teeth and gums while the pH is similar to that of plasma. Hence the benefit of sugar-free chewing gum usage.

21 a **False.** Acid-secreting cells are found only in the fundus and body. The cardia contains many mucous cells.

b **False.** Parietal cells are found in the isthmus and neck regions of the gastric glands (oxyntic glands). Chief and endocrine cells are mainly found in the basal areas.

c **True.** The muco-bicarbonate barrier protects mucosa against acids, pepsin, and ingested material (alcohol, NSAIDS, aspirin).

d **True.** Thus the mucous membrane is protected even when the pH of the gastric lumen is < 1.0.

e True The 'full' stomach may contain 2 litres of fluid.

22 a **True.** The cephalic phase lasts about 1 hour. However, the gastric and intestinal phases last up to 4 hours.

b **True.** However, 'psychic juice' is not an important mechanism in humans.

c **True.** Vagus stimulates parietal cells, chief cells, and antral gastrin cells.

d **False.** Amino acids and peptides are strong secretagogues both in the stomach and the duodenum. They act via gastrin and possibly via other GIT hormones.

e **False.** Distension strongly stimulates gastric secretion (via vagovagal reflexes and local cholinergic mechanisms).

23 a **True.** Pure parietal juice is is 9 parts HCl and 1 part KCl.

b **True.** Non-parietal juice is secreted at a rate of 1.0 ml/h and contains 0.5–2.5 mmol HCO_3^-/litre.

c **False.** K^+ is present in both parietal and non-parietal juice.

d **True.** Acid secretion rises from a basal rate of 2.5 mmol H^+/h to 25 mmol H^+/h during maximal activity.

e **False.** A gastrinoma is more likely to appear in the pancreas. Up to 100 mmolH^+/h is secreted in the Z–E syndrome.

24 a **True.** After stimulation the microvesicles coalesce, dramatically increasing the area of the apical (luminal) membrane.

b **False.** Histamine is produced, stored, and released by local mast cells and local enterochromaffin cells.

c **True.** Histamine stimulates cAMP-dependent protein kinases which then activate the proton pump (H^+,K^+-ATPase).

d **False.** PGs, especially of E class, inhibit adenylate cyclase.

e **True.** Note too that histamine has an independent synergic effect with gastrin and ACh.

25 a **False.** The hydrogen pump is on the apical (luminal) membrane only.

b **False.** H^+ is made available by intracellular reactions that involve carbonic anhydrase (CA).

c **False.** Cl^- enters the cell in exchange for HCO_3^- via an ion-exchange protein in basolateral walls; exits through luminal wall (Cl- permease) to accompany pump-extruded H^+ (see 19a).

d **True.** H^+ pump pulls luminal K^+ into cell. There is some net KCl secretion.

e **False.** Parietal cell CA is relatively unaffected by acetazolamide. (A different variety of CA is found in the kidney.)

26 a **False.** Antimuscarinic drugs reduce acid output after food by at least 30%.

b **True.** However, somatostatin (octreotide) is expensive and is a parenteral agent. Note that octreotide is a powerful vasoconstrictor of the splanchnic circulation.

c **False.** Because of the interdependency of pathways, H_2 receptor antagonists can reduce cholinergic secretion.

d **True.** A low pH also inactivates benzimidazoles.

e **True.** This is because omeprazole acts on the final common point of acid secretion. Note that omeprazole has little effect on H^+ secretion in the kidney.

27 a **False.** Ultimately acid erodes the mucosa.

b **False.** There are conflicting reports. The statement is true in Z–E syndrome. Males usually have a larger acid cell mass than females.

c **True.** The most potent stimulus to gastrin secretion is a rise in luminal pH. Acidity inhibits gastrin secretion.

d **True.** NSAIDs are the second most common cause of peptic ulcer.

e **True.** *H. pylori* contains urease. This breaks down urea before it is absorbed. Triple therapy, e.g. omeprazole, amoxicillin, and metronidazole, for 1–2 weeks heals ulcers and eradicates the bacterium in over 90% of cases.

Note that this increases the possibility of involvement of *H. jejuni* in ulceration lower down.

28   a   **True.** Ductal secretion is very important and copious.
     b   **True.** Secretin juice contrasts with CCK and vagal juice which is rich in proenzymes.
     c   **False.** Most HCO_3^- comes from the ducts by a HCO_3^- /Cl^- antiport system.
     d   **True.** The proteins produced by the acini are mainly digestive proenzymes for fat, carbohydrate, protein, and nucleic acid.
     e   **False.** Acini produce a trypsin inhibitor. This prevents activation of a cascade of pancreatic proenzymes.

29   a   **True.** Cephalic phase (vagal) and gastric phase (distension) mediate a small, predominantly acinar, secretion.
     b   **False.** H^+ in duodenum releases mucosal secretin. This is a major stimulus to pancreatic ductal water and HCO_3^- secretion.
     c   **True.** Fat acts as a secretagogue via CCK and a vagovagal enteropancreatic reflex.
     d   **True.** In the Z–E syndrome the H^+ load to the duodenum exceeds the neutralizing capacity of the duodenal–pancreatic–biliary system. Unbuffered H^+ is then free to inactivate pancreatic lipase.
     e   **True.** Hypocalcaemia accompanies acute pancreatitis in 25% of cases. This may be due to intraperitoneal saponification of calcium fatty acids in areas of fat necrosis.

30   a   **True.** Bile volume is similar to the volume of saliva secreted each day.
     b   **True.** Two-thirds of the bile is secreted by epithelial cells lining the biliary system.
     c   **False.** Bile salts control hepatocyte bile production only.
     d   **True.** Non bile acid dependent bile is chiefly controlled by secretin and CCK.
     e   **False.** The liver forms only about 0.5 g of primary bile acids per day.

31   a   **False.** Primary bile acids are conjugated with glycine and taurine in the hepatocyte prior to secretion.
     b   **True.** Most of the Na^+, Cl^-, HCO_3^-, and water is of duct origin.
     c   **True.** A normal gall bladder can only hold up to about 60 ml.
     d   **False.** Bile salts exist as macromolecular micelles which trap otherwise osmotically active material. Gall bladder (GB) bile is therefore isotonic.
     e   **True.** GB bile is more acid than hepatic bile because of HCO_3^- reabsorption in the gall bladder.

32   a   **True.** In this way the micelle becomes water soluble.
     b   **False.** Water reabsorption follows the active expulsion of Na^+ from the basolateral walls of the lining cells.
     c   **True.** 90–95% of bile salts are reabsorbed here (enterohepatic circulation).

d **False.** One loses 0.5 g of bile salts per day. This is replaced by *de novo* hepatic synthesis.

e **True.** 4–5 g of bile salts enter in a single digestive period, yet the bile salt 'pool' = 3.5 g, demonstrating the rapid recycling of the enterohepatic circulation.

33 a **False.** Cholesterol and bilirubin stones are metabolic in origin.

b **True.** An increase of lecithin increases the solubility of cholesterol.

c **True.** Oral bile salts (1–2 g/day) decrease the cholesterol:bile salts ratio. Thus bile salts help to prevent and even dissolve cholesterol stones.

d **True.** Hepatic bile is supersaturated with $CaCO_3^-$. If HCO_3^- is not reabsorbed, $CaCO_3$ may precipitate.

e **True.** The most saturated bile is formed when small volumes are secreted (night-time, between meals).

Gut hormones (Qs 34–39)

34 Gastrin:

a Gastrin stimulates degranulation of the chief cells in the stomach.

b Microvilli of G cells are in direct contact with luminal contents.

c G cells are exempt from neural control.

d Gastrin is released into lumen, absorbed, and then reaches target cells via the bloodstream.

e The effects of G17 are longer lasting than those of G34.

35 Gastrin and gastric inhibitory peptide (GIP):

a amine precursor uptake and decarboxylation (APUD) cells produce gastrin but not GIP.

b Gastrin is important in maintaining gastric mucosal growth.

c GIP comes mainly from the antrum of the stomach.

d GIP release is stimulated by distension of the stomach.

e The major effect of GIP is to inhibit gastrin effects.

36 Secretin:

a inhibits gastrin-mediated H^+ secretion in the stomach.

b works in concert with VIP on the pancreatic duct cell.

c and most other endocrine pancreatic secretagogues work via cAMP.

d is secreted in response to H^+ and fatty acids in the small gut.

e is one of the most recently discovered gut hormones.

37 Cholecystokinin (CCK):

a is released chiefly in response to fatty acids in the intestine.

b promotes the concentration of bile in the gall bladder.

c is a partial agonist of gastrin.

d potentiates secretin-stimulated release of ductal bicarbonate in the pancreas.

e stimulates appetite.

38 Vasoactive intestinal polypeptide (VIP) and somatostatin:

a VIP inhibits gastric activity.

b VIP is secretomotor to the small gut.

c VIP is a vasodilator.

d Somatostatin is found in the alpha cells of the pancreas.

e Somatostatin can be used to good effect in treating a vipoma.

39 In the digestion of proteins:

a there is marked impairment in the absence of pepsin.

b around 90% normally takes place in the jejunum.

c Pancreatic enzymes break down proteins into amino acids.

d Final digestion of polypeptides occurs at the duodenal–jejunal brush border.

e There is a single special carrier system capable of transporting all dietary amio acids into the enterocyte.

Gut hormones (Answers)

34 a **True.** Gastrin also stimulates parietal cells, mucous cells, and pancreatic acini.

b **True.** G cells are stimulated by luminal contents of stomach, mainly by peptides and amino acids.

c **False.** Non-cholinergic vagal fibres secrete gastrin releasing peptide.

d **False.** Gastrin granules discharge into local capillaries on the abluminal side of G cells.

e **False.** G17 (mainly from the antrum) is more potent but shorter acting than G34 (mainly from the intestine).

35 a **False.** APUD cells (neural crest origin) produce all gut peptide hormones.

b **True.** Gastrin is essential for stimulation of replacement cells for those sloughed off in normal wear and tear.

c **False.** GIP is produced mainly in jejunum, duodenum, and upper ileum.

d **False.** GIP is produced in response to glucose and to a lesser extent to fats in the small gut.

e **False.** The main effect of GIP is to stimulate insulin release.

36 a **True.** Secretin also inhibits gastric motility.

b **True.** Secretin and VIP both work on pancreatic duct cells via cyclic AMP mechanism.

c **False.** CCK, ACh, bombesin, and substance P all work via the inositol/calcium system.

d **True.** Secretin is the most important hormonal regulator of H⁺ in the small gut.

e **False.** Secretin was the first hormone identified (Bayliss and Starling, 1902).

37 a **True.** Fatty acids, H⁺, peptides, and amino acids in the small intestine all provoke CCK secretion.

b **False.** CCK stimulates contractions of the gall bladder and relaxes the sphincter of Oddi.

c **True.** Gastrin and CCK have identical structures in the last five amino acids at the NH₂ terminus.

d **True.** CCK potentiates a secretin effect on pancreatic ducts to a minor degree. Note the major effect of CCK and the vagus is stimulation of a pancreatic acinar secretion rich in proenzymes.

e **False.** CCK probably inhibits appetite by entering the area and so affecting nearby appetite centres.

38 a **True.** VIP is a minor inhibitor of gastric activity. It helps notably in relaxing the lower oesophageal sphincter.

b **True.** Watery diarrhea is typical of a vipoma.

c **True.** VIP is a strong vasodilator when released in the salivary glands and in penile tissue.

d **False.** Somatostatin is found in the D cells of the pancreatic islets, also in the antral, duodenal, and pancreatic mucosa, and in the hypothalamus.

e **True.** Somatostatin ('octreotide' is a synthetic analogue) is useful in treating a vipoma, in pancreatic fistulae, in peptic ulcers, and in bleeding oesophagal varices. SS inhibits the release of most gut hormones.

39 a **False.** Protein can be completely digested by enzymes in the small intestine.

b **True.** Less than 10% of protein digestion takes place in the stomach and ileum.

c **True.** Endopeptidases (trypsin, chymotrypsin, elastase) and exopeptidases (carboxypeptidases) form peptides and amino acids.

d **True.** Final protein digestion occurs in duodenum/jejunum.

e **False.** Separate systems handle the absorption of neutral, basic, dicarboxylic, and imino acids (proline, hydroxyproline).

Absorption and digestion (Qs 40–49)

40 In protein absorption:

a Di- and tripeptides can be absorbed readily into the enterocyte.

b In protein absorption only amino acids can enter the portal capillaries from the enterocytes.

c Young infants can absorb intact proteins.

d Impairment is indexed by increased fecal nitrogen.

e Pancreatic exocrine function must be less than 10% before protein malabsorption occurs.

41 Regarding protein absorption:

a Normally more than 80% of the protein in the lumen of the GIT is dietary in origin.
b In cystic fibrosis, digestion of protein products is impaired.
c The protein malnutrition in coeliac disease is essentially due to an inability to digest proteins.
d Selective GIT malabsorption of certain amino acids may be shared by the kidney.
e Amino acid absorption is virtually unimpaired in a secretory diarrhea.

42 In carbohydrate digestion and absorption:

a Pancreatic amylase rapidly hydrolyses maltose.
b Salivary amylase is very important in the newborn.
c Both salivary and pancreatic amylase attack 1,4-glycosidic linkages.
d Disaccharidases are secreted by glands in the jejunal mucosa.
e Up to half of the ingested glucose is absorbed in the first 20 cm of the jejunum.

43 In carbohydrate absorption:

a Absorption only occurs after carbohydrates have been broken down to hexose sugars.
b Hexose sugars share a cotransport system with amino acids.
c D-glucose enters enterocytes quicker than L-glucose.
d Non-sodium-dependent hexose transport is relatively small compared to sodium-dependent transport.
e Sodium cotransport is an example of primary active transport.

44 In the digestion of carbohydrate:

a Lactose intolerance is due to a failure to absorb the lactose molecule.
b A child with lactose intolerance has usually had the condition since birth.
c Absence of the Na^+/glucose transporter in the intestine has little clinical significance.
d Fructose is transported along with glucose via the Na^+/glucose cotransport system.
e Lithium might be expected to increase hexose absorption.

45 In the digestion of fats:

a Sodium taurocholate can emulsify dietary fat droplets in the duodenum.
b Pancreatic trypsin speeds fat digestion.
c Pancreatic lipase breaks up the lipid emulsion into water-soluble micelles.
d Colipase splits off monoglycerides (MGs) and fatty acids (FAs) from emulsified fat.

e   A small number of water-soluble micelles can be absorbed directly by the enterocyte.

46 In fat absorption:

a   All FAs and MGs are reconstituted to triglyceride in the mucosa of the small intestine.
b   Short- and medium-chain fatty acids are water soluble.
c   A chylomicron consists of cholesterol and phospholipid covered in a layer of protein only.
d   Chylomicrons travel via lacteals to the liver.
e   Fat-soluble vitamins are absorbed when solubilized by cholesterol.

47 In fat malabsorption:

a   There should be no neutral fat in the faeces of a normal adult.
b   Fats can be degraded appreciably in the colon.
c   Fatty acids can be absorbed in the colon.
d   Calcium oxalate renal stones may result from fat malabsorption.
e   Prepancreatic lipase (saliva, stomach) can, in time, make up for the loss of pancreatic lipase.

48 Sodium absorption:

a   About 60 g of sodium are absorbed each day (1 mol = 58 g).
b Na^+ is cotransported with glucose and amino acids in the colon.
c   Aldosterone influences Na^+ absorption in colon.
d   90% of Na^+ absorption occurs in the small intestine.
e   Less than a gram of sodium appears in normal stools each day.

49 Chloride in the gut:

a Cl^- is absorbed following the electrical gradient created by the active absorption of Na^+.
b   Most chloride absorption in the jejunum is by the paracellular route.
c   The Cl^- channel is found in both small intestine and colon.
d   The Cl- channel couples the exchange of $2Cl^-$ for one K^+ and one Na^+ ion.
e   Normally cAMP inhibits the chloride channel.

Absorption and digestion (Answers)

40   a   **True.** Once small peptides are inside the enterocyte they are broken down by peptidases into amino acids which then enter the portal system.
b   **False.** Some peptides resist hydrolysis and enter the portal blood intact, e.g. TRH (thyrotrophin releasing hormone, a tripeptide) and some peptides from gelatin.
c   **True.** In the first 36 h of life neonates can absorb IgG by endocytosis (achlorhydria aids this).

    d **True.** Increased fecal nitrogen is called creatorrhea.

    e **True.** Protein malabsorption results in a putrefactive diarrhea. Enteric flora degrade undigested protein.

41 a **False.** The origin of luminal protein is 25% from sloughed mucosal cells, 25% from secreted enzymes, and 50% from ingested protein.

    b **True.** In cystic fibrosis there is a blockage/abnormality of pancreatic secretions. DNA from recruited inflammatory cells is a significant factor in the raised viscosity of secretions in the GIT and the lung. Recombinant DNAase is clinically useful and synthetic replacement more so.

    c **False.** Coeliacs have a gluten-sensitive enteropathy in which the mucosal damage causes a defective absorption of protein, carbohydrate (CHO), fat, Fe, and vitamins. Note that oats are a good alternative to wheat. (Oats do not share the antigenic gliadin component of wheat.)

    d **True.** Neutral amino acids in Hartnup disease are an example.

    e **False.** One's capacity to reabsorb amino acids is overwhelmed in secretory diarrheas.

42 a **False.** Pancreatic amylase generates tri- and disaccharides from starch.

    b **True.** Salivary amylase is important in the neonate because because pancreatic amylase is insufficient at this time.

    c **True.** Apart from infancy, pancreatic amylase is much more important than salivary amylase.

    d **False.** Disaccharidases are present in the brush border of the small intestine.

    e **False.** Normally all, or virtually all, glucose absorption is completed in the proximal jejunum.

43 a **False.** Pentose sugars are readily absorbed (Na^+ cotransport).

    b **True.** Many amino acids compete with hexoses for a common Na^+-dependent cotransport entry into enterocytes.

    c **True.** Stereospecificity is displayed by transport systems.

    d **False.** Non-sodium-dependent transport normally accounts for about 50% of hexose absorption.

    e **False.** Sodium/glucose cotransport is driven by the Na^+ gradient that has been actively created by Na^+,K^+-ATPase in the basolateral walls of the enterocyte, i.e. this is a secondary active transport process.

44 a **False.** Lactose intolerance is caused by a failure to digest lactose. Intact lactose is not absorbed by anyone.

    b **False.** Lactase is rarely absent at birth. Levels decline after 2–3 years (especially in non-Caucasians). When lactose intake is limited to the equivalent of 240 ml of milk/day symptoms previously attributed to lactose intolerance usually disappear.

    c **False.** Glucose/galactose malabsorption occurs in the absence of the Na^+/glucose transporter. This causes severe diarrhea.

d **False.** Fructose absorption is independent of glucose and galactose. Fructose enters the enterocyte via GLUT 5 and leaves via GLUT 2.

e **False.** Li^+ displaces Na^+ and thus reduces hexose cotransport.

45 a **False.** Micellar bile salts–phospholipid–cholesterol emulsify by attaching to droplets via their phospholipid–cholesterol moiety. On its own Na taurocholate cannot emulsify.

b **True.** Pancreatic trypsin activates procolipase, prophospholipase, and procarboxyl esterase.

c **True.** Final micelles are complexes of bile salts, cholesterol, mono-glycerides, and fatty acids. Fat-soluble parts are in the centre. Panceatic lipase acts on fats that have been emulsified, releasing free fatty acids and 2-monoglycerides.

d **False.** Colipase anchors lipase to fat droplets.

e **False.** Passive diffusion of MGs, cholesterol, and FAs occurs from the micelles in the unstirred layer into mucosal cells.

46 a **False.** Short-chain fatty acids (<10–12 carbons) go directly to the liver via the portal vein.

b **False.** Fats are not broken down in the colon apart from minor bacterial lipase action.

c **False.** The chylomicron centre also contains triglycerides and often trace lipids.

d **False.** Chylomicrons travel via lacteals to the thoracic duct and thence to the systemic venous system.

e **False.** Fat-soluble vitamins must be solubilized by bile acids. They are absorbed from micelles.

47 a **False.** Normally faecal fat is up to 6 g /day, or less than 6% of ingested fat.

b **False.** Except for minor bacterial lipase action.

c **False.** Any fatty acid that enters the colon is excreted in the faeces.

d **True.** Dietary oxalate that persists in solution in the small gut is absorbed in the colon where FAs increase paracellular absorption.

e **False.** If pancreatic secretion is < 10% normal, steatorrhea occurs.

48 a **True.** Only 10–15 g of sodium comes from the diet. The rest is from GIT secretions.

b **False.** Na^+/glucose cotransport occurs in the small intestine.

c **True.** Aldosterone affects the colon where K^+ and H^+ are exchanged for Na^+.

d **True.** Most sodium is absorbed in the small gut where it is cotransported with hexoses and amino acids or exchanged for H^+ across the apical mem-brane, or diffuses in by the paracellular route. The basolateral Na^+ pump is essential in all cases.

e **True.** About 5–10 mmol Na^+ is lost daily (0.27–0.58 g/day).

49 a **True.** Cl⁻ is also absorbed via HCO_3^-/Cl^- exchange in ileum and colon.

b **True.** There is no apical membrane HCO_3^-/Cl^- exchange in the jejunum.

c **True.** But the chloride channel is particulary abundant in the basal crypt cells of villi.

d **False.** $2Cl^-$, $1Na^+$, and $1K^+$ are cotransported into the cell on the wall opposite to that containing the Cl⁻ channel.

e **False.** Both intracellular cAMP and Ca^{2+} open up the Cl⁻ channel. Cholera toxin switches on cAMP, producing an efflux of Cl⁻ resulting in the characteristic watery stool.

Diarrhea and control of food intake (Qs 50–52)

50 The following statements about diarrhea are correct:

a Diarrhea is usually due to problems originating in the large intestine.

b Magnesium inhibits intestinal movements.

c A missing HCO_3^-/Cl^- exchanger in the colon leads to diarrhea with a high HCO_3^- content.

d The exotoxin of *V. cholerae* stimulates G proteins in the epithelial cells of the GIT.

e Serotonin may cause diarrhea by increasing cytoplasmic calcium.

51 In the management of diarrhea:

a standard oral rehydration therapy (ORT) can dramatically reduce the frequency of stools.

b a low-osmolality ORT can reduce stool output significantly.

c the sodium:sugar ratio should be close to unity in any ORT one uses.

d Vitamin A supplementation helps to prevent diarrhea in young children.

e Zinc contributes to diarrhea in young children.

52 About the control of food intake:

a The satiety centre is located in the ventromedial hypothalamus.

b The hunger or feeding centre is located in the medulla oblongata.

c The quality of food that is eaten is partly determined by the hypothalamus.

d A high level of blood glucose stimulates the satiety centre and a low level of glucose inhibits it (glucostatic theory).

e The lipostatic theory states that feeding is under the long-term control of total body adipose mass.

Diarrhea and control of food intake (Answers)

50 a **False.** Most cases of diarrhea originate from problems in the small intestine.

b **False.** Mg is poorly absorbed. Each mmol holds up 10 ml faecal water. Mg-induced diarrhea is common.

c **False.** It leads to diarrhea with a high Cl⁻ content. Metabolic alkalosis is from retention of HCO_3^-.

d **True.** G proteins cause increase in cAMP which opens Cl⁻ channels and causes a secretory diarrhea.

e **True.** Serotonin may be partly responsible for diarrhea or intussusception after overeating bananas (rich in 5-HT).

51 a **False.** The chief purpose of ORT is to replace fluid and electrolytes which can be lost fatally quickly.

b **True.** Solutions containing 60 mmol Na and 84 mmol glucose/l (osmolality of 224 mosm/kg) can reduce stool output by 28%.

c **True.** Higher ratios may cause hypertonicity and exacerbate diarrhea.

d **True.** Vitamin A helps maintain epithelial surfaces. Note that excessive amounts of vitamin A but not provitamin A will cause birth defects.

e **False.** Zinc deficiency is assoicated with chronic diarrhea which may be due to a failure in the immune system.

52 a **True.** The satiety centre is probably not confined to the hypothalamus.

b **False.** The hunger centre is found in the ventrolateral hypothalamus.

c **False.** The quality of food ingested is probably determined by the cortical areas of the limbic system.

d **True.** Satiety cells are also more active if their glucose utilization is high (i.e. A-V glucose difference across them high).

e **True.** The total body adipose mass probably monitors the set-point for body weight rather than food intake *per se*. The set-point probably responds to leptin and the average concentration of plasma free fatty acids. The adipostat is almost certainly not confined to the hypothalamus. Note that leptin is a protein that regulates body weight and that is encoded in adipocytes by the *ob* gene.

8

Hormones

Cell signalling (Qs 1–4)

1 *In the transduction of chemical signals via membrane receptors:*

a   Thyroid hormone signals its message via a plasma membrane receptor.

b   Steroid hormones attach to a specific plasma membrane receptor.

c   A ligand is a membrane protein that combines with some specific substance not necessarily protein in nature.

d   Insulin signals its message via second messenger cAMP.

e   Ligand-gated ion channels (L-channels) are regulated by protein kinases (PKs).

2 *With chemical signalling via membrane receptors:*

a   Ion-channel phosphorylation is confined to cation channels.

b   Kinase phosphorylation of channels is abnormal in cystic fibrosis (CF).

c   Second messengers are free cytoplasmic substances that activate a variety of protein kinases.

d   G (guanine nucleotide binding) proteins are important cytoplasmic second messengers.

e   The cAMP pathway always involves an increase in cytoplasmic cAMP.

3 *Second messengers:*

a   Phosphodiesterase (PDE) converts cAMP to AMP + inorganic phosphate.

b   PDE inhibitors augment the effects of cAMP by slowing down its degradation.

c   The cGMP pathway is at least as widespread as the cAMP pathway.

d   Phosphatidyl inositol bisphosphate (PIP_2) is a second messenger.

e IP_3/Ca^{2+} is a very widespread chemical transduction system.

4 *Intracellular signalling:*

a   Inositol is a lipid.

b   Inositol normally occurs in a bound form in the body.

c   Cortisol combines with a specific protein receptor on an intracellular organelle.

d   G proteins couple cell surface receptors with intracellular effectors.

e   Increased G protein activity is typical of generalized resistance to hormone action, e.g.resistance to PTH (parathyroid hormone), thyrotrophin, and gonadotrophins.

Cell signalling (Answers)

1   a   **False.** Thyroid hormone enters cells and then signals its message.

b   **False.** Steroids (lipid soluble) pass through plasma membrane and attach to receptors on intracellular organelles.

c **False.** A ligand is anything that combines with a receptor whether the receptor is on a membrane or in the cell.

d **False.** The insulin–receptor complex exerts its intracellular effect by directly phosphorylating cytoplasmic proteins.

e **True.** PKs on the inner surface of the plasma membrane phosphorylate subunits of different ion channels. Note that the term L-type channel is used to signify long-lasting and large, as well as ligand-gated, channels.

2 a **False.** Ion-channel phosphorylation also applies to anion channels, e.g. Cl^- channel.

b **True.** Cl^- channel phosphorylation is abnormal in CF and in some cases of myotonia congenita.

c **True.** The most important second messengers are cAMP, cGMP, and Ca^{2+}.

d **False.** G proteins are a family of intrinsic membrane proteins. Gs stimulates and Gi inhibits activity.

e **False.** Activation of α-receptors (amines) and D_2 receptors (dopamine) causes a decrease in cAMP.

3 a **True.** PDE limits the effects of raised cytoplasmic cAMP.

b **True.** PDE inhibitors, e.g. caffeine, theophylline, and milrinone, augment cAMP effects.

c **False.** However, cGMP is stimulated notably by atrial natriuretic factor and nitric oxide.

d **False.** PIP_2 is membrane bound. G protein activation releases inositol trisphosphate (IP_3) and diacylglycerol (DAG) into the cell.

e **True.** IP_3/Ca^{2+} is important in Ca^{2+}-dependent reactions (muscle contraction, secretion), in phosphorylation (liver) and in sensory perception and neuromodulation.

4 a **False.** Inositol has the same empirical formula as glucose, $C_6H_{12}O_6$.

b **True.** Inositol occurs as phosphatidyl inositol (PI) from which PI-bisphosphate and PI-trisphosphate are actively synthesized in the plasma membrane.

c **True.** Protein receptors for cortisol are highly specific.

d **True.** G proteins act as on-off switches for cellular signalling.

e **False.** Most patients with general hormone resistance have a 50% reduction of the G proteins that stimulate adenylate cyclase (which is needed by so many hormones).

Non-pituitary hormones (Qs 5–44)

5 *About the thyroid gland:*

a The thyroid is derived from ectoderm.

b The thyroid has a very rich sympathetic supply.

c Thyroid stimulating hormone (TSH) interacts with follicular cell membrane receptors.

d TSH stimulates mast cells lying between follicular cells.

e The size of the gland is largely determined by the physical activity of the subject.

6 Concerning iodine and the thyroid:

a Iodine is converted to iodide before absorption in GIT.

b Iodide trapping is helped by an electrical gradient into the follicle cell.

c Iodide trapping is a primary active process.

d TSH stimulates iodide uptake at all iodide trapping sites in the body.

e Perchlorate is a competitive inhibitor for iodide trapping.

7 In the formation and secretion of thyroid hormone:

a Thyroglobulin (TGL) is synthesized and stored in the colloid spaces found in the thyroid gland.

b TGL contains MIT and DIT (mono- and di-iodotyrosine).

c A high iodine intake stimulates incorporation of iodine into TGL.

d Thyroid hormone is secreted by exocytosis from colloid into the ECF.

e Thyroid secretions have a T_3:T_4 ratio of about 9:1.

8 Regarding thyroid hormones and their availability:

a Free T_3 is the overwhelmingly active form of thyroid hormone.

b The T_3 receptor is present in the nucleus of target cells.

c Albumin is the plasma protein with the highest binding affinity for T_3 and T_4.

d Thyroid-binding globulin (TBG) is increased in pregnancy.

e Fluctuations of thyroid hormone binding is an important cause of clinical dysfunction.

9 The following points are true about about T_3, thyroid disease:

a Correction of iodine deficiency in the third trimester prevents cretinism.

b T_3 stimulates protein synthesis.

c T_3 is an important stimulator of Na^+,K^+-ATPase.

d Dysfunction of T_3 receptors may cause attention deficit hyperactivity disorder (ADHD) in children.

e Smoking has a marked association with Graves ophthalmopathy.

10 In the hormonal control of calcium:

a Calcitonin comes from cells of neuroectodermal origin.

b Calcitonin is an important physiological regulator of plasma calcium.

c Hypocalcaemia is the main stimulus for calcitonin release.

d Parathyroid hormone (PTH) is essential for life.

e PTH is stored in granules in the parathyroid.

11 *In the control of calcium and phosphorus:*
a   PTH enhances PO_4^- reabsorption in the proximal renal tubule.
b   PTH promotes the activation of vitamin D in the liver to 1,25 OH D_3.
c   PTH has a negative feedback effect on the anterior pituitary.
d   PTH mobilizes phosphate from bone, thereby increasing plasma PO_4^-.
e   Neither PTH nor calcitonin can control acute hyperphosphataemia.

12 *Atrial natriuretic factor (ANF):*
a   type A is chiefly found in the myocytes of the right atrium.
b   type A is released in hypovolaemia.
c   inhibits the renin–angiotensin–aldosterone axis.
d   causes natriuresis and vasodilation.
e   has its actions potentiated by neutral endopeptidase inhibitors.

13 *About melatonin:*
a   It is derived from the amino acid tyrosine.
b   Hydroxyindole-O-methyltransferase (HIOMT) is necessary for its synthesis.
c   Melatonin induces ovulation.
d   Melatonin almost certainly entrains the body's 24-hour 'clock'.
e   Melatonin can have an hypnotic effect.

14 *In their 24-hour circadian rhythms:*
a   Melatonin is highest during the night.
b   Prolactin is normally highest during the day.
c   Growth hormone (GH) levels are highest at night.
d   Cortisol is highest around dawn.
e   Body temperature is highest at night.

15 *The adrenal medulla:*
a   is relatively well developed at birth.
b   is called the organ of Zuckerkandl before birth.
c   produces adrenaline (A) from noradrenaline (NA).
d   is largely of neural crest origin.
e   contains chromaffin cells, most of which are capable of secreting both A and NA.

16 *Adrenal medulla function:*
a   It normally secretes about 80% adrenaline and 20% noradrenaline.
b   Amines are stored in granules in chromaffin cells.
c   Amines are stored, and secreted with, a glycoprotein, chromogranin A.
d   Secretion is triggered by postganglionic sympathetic terminals.
e   ACh hyperpolarizes chromaffin cells.

17 *In the synthesis and degradation of medullary amines:*
a   Tyrosine is the precursor of adrenaline (A) and noradrenaline (NA).

b   Cortisol stimulates dopamine β-hydroxylase (dopamine $\Rightarrow$ NA).
c   Cortisol stimulates methylation of noradrenaline to adrenaline.
d   Catecholamines have a long half-life in extraneuronal tissue.
e   End products of amine metabolism eventually appear in the urine.

18 In the normal function of the adrenal medulla:

a   The basal secretion of catecholamines is important for maintaining vascular tone.
b   Adrenaline causes vasoconstriction in the skin.
c   Basal NA secretion has a significant biological effect.
d   Significant secretion is only stimulated by sympathetic activity.
e   Adrenaline inhibits lipase in adipose tissue.

19 The following is true of cateclolamines:

a   Adrenaline stimulates the release of insulin.
b   Adrenaline increases phosphorylase activation in the liver.
c   Noradrenaline increases the force and rate of cardiac contraction.
d   Noradrenaline can result in a bradycardia.
e   Adrenaline has more potent metabolic actions than noradrenaline.

20 Adrenoreceptors:

a β1- and β2-adrenoreceptors stimulate adenyl cyclase.
b α1-receptors are on the presynaptic membrane.
c α2-receptors work by inhibiting adenyl cyclase.
d   Cortisol enhances activity at β-adrenoreceptors.
e   In normal people, tissue density of β-adrenoreceptors is upregulated by continuous stimulation.

21 During the synthesis of steroid hormones in the adrenal cortex:

a   Steroids are synthesized *de novo* from cholesterol and acetate.
b   Cholesterol comes from circulating low density lipoprotein (LDL).
c   In acute stimulation, steroidogenesis is mainly from circulating LDL.
d   Cytochrome P450 cleavage enzyme is critical for steroidogenesis.
e   Steroid hormones are fully assembled in the mitochondria.

22 Adrenal cortical hormones:

a   Secretion of adrenal androgens is controlled by gonadotrophins.
b   ACTH is greatly elevated in the adrenogenital syndrome.
c   Adrenogenital syndrome is associated with virilization.
d   Congenital adrenal hyperplasia causes precocious puberty in boys.
e   Adrenal oestrogens are important as a source of oestrogen in women of reproductive years.

23 Adrenal cortical hormones:

a   Corticosteroids are stored in the adrenal cortex.

  b Adrenal androgens have a considerable biological action.
  c Cortisol is mostly bound to plasma proteins in the circulation.
  d Aldosterone is mostly bound to plasma proteins in the circulation.
  e Cortisol and aldosterone have roughly the same half-life.

24 In the regulation of adrenal cortical hormones:

  a Cortisol and aldosterone are conjugated with glucuronic acid in the liver.
  b Their half-life depends on the ease with which the liver accesses and metabolizes them.
  c Cushing symptoms are more likely in pregnancy because of the raised plasma cortisol found during pregnancy.
  d Plasma cortisol is usually grossly elevated in severe depression.
  e Cortisol feedback determines the rate of production of adrenal androgens.

25 Cortisol:

  a is the only glucocorticoid secreted in significant amounts in humans.
  b acts at cell level by directly stimulating/inhibiting enzymes.
  c increases liver glycogen stores.
  d stimulates the use of glucose in peripheral tissues.
  e helps to mineralize bone.

26 Cortisol effects include:

  a inhibition of the growth of fibroblasts.
  b stimulation of the mobilization of membrane arachidonic acid.
  c increasing circulating eosinophils.
  d a marked enhancing effect on amines and on glucagon.
  e significant anti-inflammatory and anti-immune effects only when present in excess.

27 In the control of glucocorticoids:

  a ACTH controls the release of cortisol.
  b Cortisol inhibits the release of ACTH and CRH (corticotrophin releasing hormone).
  c Cortisol is normally secreted in steady amounts during the day.
  d The ACTH rhythm is clearly related to a light/darkness pattern.
  e ACTH acts slowly (> 12 h) to promote steroidogenesis.

28 In tests of adrenocortical–hypothalamic function:

  a A hypoglycaemia of 3.3 mmol/l (60 mg/dl) is adequate to raise the blood cortisol significantly.
  b An insulin-induced hypoglycaemia directly stimulates ACTH release.
  c The CRH test (CRH 100 μg IV) gives a greater ACTH response than the insulin test.
  d The dexamethasone suppression test separates adrenal from pituitary causes of hypercortisolism.
  e Metyrapone is a strong stimulant of cortisol production.

29 *In disorders of cortisol secretion:*

a The skin pigmentation of primary Addison's disease is due to excess ACTH.

b Addison's disease usually only implicates glucocorticoids.

c Cushing's syndrome only implicates excess glucocorticoids.

d Hyperpigmentation in Cushing's disease is due to high cortisol.

e In adrenal tumours plasma cortisol and ACTH are both elevated.

30 *Concerning aldosterone formation and secretion:*

a Aldosterone is formed only in the zona glomerulosa.

b Angiotensin II is the most important regulator of aldosterone.

c The renin–angiotensin system is found in the adrenal cortex.

d ACTH has no effect on aldosterone secretion.

e Potassium stimulates aldosterone release.

31 *Aldosterone effects include:*

a the reabsorption of Na^+ in the proximal renal tubule.

b the reabsorption of Na^+ in the renal cortical collecting duct.

c an increased synthesis of Na^+,K^+-ATPase in the apical membrane of target duct epithelial cells.

d the retention of salt and water in ducts of most exocrine glands.

e increased Na^+ and K^+ conductance at target membranes.

32 *Hypoaldosteronism:*

a Salt-wasting forms of congenital adrenal hyperplasia are associated with hypoaldosteronism.

b Any form of adrenal insufficiency may be associated with hypo-aldosteronism.

c High levels of aldosterone are compatible with clinical hypoaldosteronism.

d Adrenal insufficiency is typified by hyperkalaemia, hyponatraemia, and hypertension.

e Aldosterone normally takes 10–30 min to act.

33 *Hyperaldosteronism:*

a Primary hyperaldosteronism is associated with a low plasma renin.

b Aldosterone is significantly regulated by ACTH in glucocorticoid-suppressible hyperaldosteronism (GSA).

c Liquorice causes hypertension by allowing cortisol to attach to mineralo-corticoid receptors.

d Elevated plasma cortisol can suppress renin and aldosterone secretion.

e Excess aldosterone inhibits renin secretion at renal level.

34 *The functioning of the pancreas as an endocrine organ:*

a Insulin, glucagon, somatostatin (SS), and pancreatic polypeptide (PP) are all found in different cells in the islets of Langerhans.

b Beta cells are by far the most numerous in the islets.

c Pancreatic polypeptide stimulates the exocrine pancreas.

d The main action of SS here is to inhibit insulin and stimulate glucagon release.

e Somatostatin of the pancreas differs slightly from hypothalamic SS.

35 Human insulin:

a is a large protein containing four polypeptide chains.

b is identical with that from other higher mammals.

c is complexed with copper in the storage vesicles.

d is almost all degraded in the liver.

e has a half-life of about 30 min.

36 In the control of insulin secretion:

a Blood glucose is the most powerful stimulus.

b Amino acids are inhibitory.

c Vagal stimulation increases secretion.

d Beta agonists strongly increase secretion.

e There is a larger insulin response after IV than after a similar amount oral glucose load.

37 The following statements about insulin secretion are true:

a There is no secretion of insulin between meals (stable plasma glucose and amino acid levels).

b Response to glucose reaches a peak and then gradually wanes.

c Insulin-like growth factors (IGF-I, II) are secreted with insulin.

d Glucagon inhibits the release of insulin.

e Ketone bodies and fatty acids are important stimulants of insulin secretions in starvation.

38 The insulin receptor:

a is the same as that for IGFs.

b has a similar structure to insulin.

c has tyrosine kinase on its external domain.

d is ultimately internalized with insulin.

e is down-regulated at all stages of life by insulin.

39 Insulin:

a is the most powerful hypoglycaemic agent in the body.

b inhibits free fatty acid (FFA) release from tissues.

c inhibits lipoprotein lipase (LPL).

d inhibits hormone-sensitive lipase.

e facilitates transport of amino acids secondary to glucose.

40 In the glucose transporter system:

a Glucose transporters (GLUTs) have the ability to carry glucose against its concentration gradient ('uphill').

b At least one GLUT isoform is expressed constitutively in all cells.
c Both erythrocytes and brain cells are very rich in GLUT-1.
d GLUTs only carry glucose in one direction across membranes.
e Insulin increases membrane GLUTs especially in myocytes and adipocytes.

41 *Insulin resistance:*
a affects all the actions of insulin.
b Antibodies to insulin are probably the most common cause of insulin resistance.
c is commonly caused by subcutaneous degradation of injected insulin.
d is caused by mutations of the insulin-receptor gene.
e is known to be due to a functional defect in genes that govern glucose transport or metabolism.

42 *Insulin resistance:*
a Antibodies to the insulin receptor are a well-recognized cause of insulin resistance.
b Counter-regulatory hormones cause insulin resistance.
c Obesity is associated with increased insulin sensitivity.
d Increased resistance is often found in normal physiological states.
e Sensitivity to insulin is increased in stressful states.

43 *Insulin facilitates glucose uptake:*
a in the liver.
b in the brain.
c in erythrocytes.
d in the intestine.
e in the crystalline lens of the eye.

44 *Glucagon:*
a consists of a double polypeptide chain similar to insulin.
b stimulates the secretion of somatostatin.
c stimulates gluconeogenesis at physiological levels.
d is the most important hormone in preventing fasting hypoglycaemia.
e is a strong lipolytic agent in adipose tissue.

Non-pituitary hormones (Answers)

5 a **False.** The thyroid gland is derived from the endoderm of the cephalic part of the GIT.
b **True.** The sympathetic supply to the thyroid is vasomotor and also supplies individual follicular cells.
c **True.** TSH receptors activate adenylate cyclase and so excite all metabolic processes of the cell.
d **True.** Mast cells release amines and histamine and so increase local blood flow and the metabolism of follicular cells.
e **False.** TSH is the main determinant of the size of the thyroid.

6   a   **True.** 150 μg iodine needed/day. Goitre is common if intake < 60 μg/day.
    b   **False.** The inside of a follicular cell is −50 mV. This repels the entry of the plasma iodide anion.
    c   **False.** Energy for iodide pumping is provided by the sodium pump.
    d   **False.** TSH has no effect on iodide traps in salivary glands, breast, placenta, ciliary body, or gastric mucosa.
    e   **True.** Perchlorate and periodate are competitive inhibitors of I⁻ at the thyroid gland.

7   a   **False.** Synthesis of TGL occurs in follicular cells.
    b   **True.** TGL contains MIT, DIT, T_3 (MIT+DIT) and T_4 (DIT+DIT).
    c   **False.** A large intake of iodine inhibits its organification (the Wolff–Chaikoff effect).
    d   **False.** Colloid is reabsorbed into the follicular cell, lysozyme proteases digest it and then T_4 and T_3 are secreted into the ECF.
    e   **False.** The reverse is true; T_3:T_4 = 1:9. About 80 μg of T_4 is secreted/day.

8   a   **True.** Most T_3 is formed by de-iodination of T_4 in liver, kidney, and target cells.
    b   **True.** T_4 is first converted to T_3 in the target cell cytoplasm.
    c   **False.** Thyroid-binding globulin and prealbumin have greater affinity for T_3 and T_4. Note they are present in lower concentration than albumin.
    d   **True.** Oestrogens stimulate the hepatic synthesis of TBG.
    e   **False.** Free hormone levels remain the same. Altered free T_3/T_4 influence TSH appropriately.

9   a   **False.** The best results occur if iodine deficiency is corrected before pregnancy.
    b   **True.** T_3 synergizes with GnRH, GH, and somatomedins (in both their production and systemic actions).
    c   **True.** The stimulatory effect of T_3 on the sodium pump probably underlies the calorigenic and metabolic stimulating effects of T_3.
    d   **True.** Resistance to T_3 is caused by a mutation in receptor gene. The absence of T_3 causes mental retardation.
    e   **True.** Smokers have a 7.7 times greater chance of severe eye problems in hyperthyroidism than non-smokers.

10 a   **True.** Parafollicular C cells originate from ultimobranchial glands. Found mainly in the thyroid, some in the thymus.
    b   **False.** Changes in PTH secretion is the main control of serum Ca^{2+}.
    c   **False.** Hypercalcaemia causes the release of calcitonin. Gastrin, CCK, and glucagon are less important secretagogues.
    d   **True.** The absence of PTH causes fatal hypocalcaemia.
    e   **False.** The biosynthesis and secretion of PTH are closely coupled.

11 a   **False.** PTH decreases PO_4 reabsorption in the PCT and enhances Ca^{2+} reabsorption in the distal tubule.

b **False.** PTH facilitates activation of vitamin D in the kidney. Active D_3 promotes calcium absorption in the GIT.

c **False.** PTH responds solely to the level of ionic plasma calcium.

d **True.** PTH is also phosphaturic. Thus it has opposing effects on the level of plasma PO_4^-.

e **True.** Endogenous phosphatonin (origin uncertain) is phosphaturic and lowers the serum PO_4^- in normal people. It has therapeutic possibilities in rhabdomyolysis and in tumour-induced osteomalacia.

12 a **True.** ANF: type B is found mainly in the ventricle; type C in the vascular endothelium.

b **False.** ANF is released by stretch (hypervolaemia, heart failure).

c **True.** ANF reduces renin release, blocks ACE and blocks aldosterone release.

d **True.** ANF decreases renin, and depresses fluid/salt drive.

e **True.** Endopeptidases have a possible role in treating heart failure, hypertension, CRF, and cor pulmonale. They may well have anti-atherosclerotic effects also.

13 a **False.** Melatonin is derived from serotonin/tryptophan.

b **True.** Light (eye–cord–sympathetic–pineal) suppresses HIOMT.

c **False.** Melatonin suppresses gonadal function. Thus it has a possible contraceptive use.

d **True.** The 'biological clock' resides in the suprachiasmatic nucleus of the hypothalamus. Many melatonin receptors are found here.

e **True.** Melatonin is hypnotic in doses of 5 mg–5 g. It is sometimes called the 'hormone of darkness'.

14 a **True.** Light inhibits HIOMT which is essential for the synthesis of melatonin.

b **False.** Prolactin is highest during the night.

c **True.** Note there is a traditional belief that a child grows when sleeping. This is perhaps based on the fact that an individual is always taller in the morning but shrinks due to disc flattening during the day.

d **True.** The cortisol rhythm is lost in cancer of the adrenal cortex.

e **False.** Body temperature is highest in the afternoon. It falls at night.

15 a **False.** Although the adrenal medulla begins to function from 10 weeks' gestation.

b **False.** This is the para-aortic body. It supplies noradrenaline for vascular tone until the adrenal medulla is properly developed.

c **True.** The adrenal medulla has a special N-methyltransferase that methylates NA to A.

d **True.** Origin of adrenal medulla is similar to that of sensory ganglia, parts of pharynx, jaw, and the squamous bones covering the forebrain.

e **False.** There are two types of chromaffin cell: one secretes A, the other NA.

16 a **True.** The adrenaline : noradrenaline ratio approaches unity and can even be reversed in shock and severe stress.

   b **True.** Granules containing amines are discharged by exocytosis.

   c **True.** This allows amine storage in an osmotically acceptable form.

   d **False.** The secretion of the adrenal medulla is triggered by by cholinergic pre-ganglionic sympathetic fibres.

   e **False.** ACh depolarizes chromaffin cells; Ca^{2+} enters and degranulation follows.

17 a **True.** Tyrosine $\rightarrow$ dihydroxyphenylalanine (DOPA) $\rightarrow$ dopamine $\rightarrow$ NA $\rightarrow$ A.

   b **False.** ACTH and sympathetic nerves stimulate NA production at this point.

   c **True.** Cortisol draining from the adrenal cortex to the medulla stimulates the methylation of noradrenaline to adrenaline.

   d **False.** Catecholamines have a half-life of about 2 min in the circulation. Catechol o-methyltransferase (COMT) degrades them in most tissues. Monoamine oxidase (MAO) degrades them at nerve terminals.

   e **True.** About 50% appear in urine in free or conjugated forms; about 35% as vanillylmandelic acid (VMA).

18 a **False.** Local neurogenically released NA, endothelial vasoactive agents, and the local environment are the important factors for the maintenance of resting vascular tone.

   b **True.** Adrenaline vasoconstricts in skin and in many viscera. It probably vasodilates in skeletal muscle.

   c **False.** NA levels need to go to 1800 pg/ml to exert haemodynamic and metabolic effects. These levels occur in stress.

   d **True.** Pain, excitement, fear, haemorrhage, and trauma all work via sympathetic system ('stress reaction').

   e **False.** Thus adrenaline increases fatty acids in plasma. It also stimulates ketogenesis in the liver. Note there is resistance of $\beta3$-receptors in adipocytes to *circulating* adrenaline.

19 a **False.** Adrenaline inhibits insulin release via $\alpha1$-receptors.

   b **True.** Thus adrenaline increases hepatic glycogenolysis and so the blood sugar is raised.

   c **True.** This is because noradrenaline stimulates $\beta1$-receptors in addition to $\alpha1$-receptors.

   d **True.** NA is a powerful vasoconstrictor. This stimulates the baroreflex by raising the systemic blood pressure which allows increased vagal discharge to the heart.

   e **True.** However, NA is a more powerful vasoconstrictor than A.

20 a **True.** There are mainly, but not exclusively, $\beta1$-receptors in the heart; $\beta2$ predominate in the in bronchi, uterus, and skeletal muscle; $\beta3$ in adipocytes, especially in brown adipose tissue.

   b **False.** $\alpha1$-receptors are postsynaptic receptors. NA strongly stimulates them.

c **True.** α2-receptors are autoregulatory. They are found on the presynaptic membrane.

d **True.** Cortisol increases cAMP at β1- and β2-receptors.

e **False.** Both β1- and β2-receptors are down-regulated by their ligands. However, β3-receptors resist such desensitization. Note there is classic down-regulation of cardiac β-receptors in chronic heart failure (tachyphylaxis).

21 a **True.** Cholesterol is stored in lipid droplets in cytoplasm.

  b **True.** LDL receptors are abundant in certain endocrine organs, e.g. adrenal cortex, ovaries, testes.

  c **False.** Cytoplasmic cholesterol is the chief source in acute steroid stimulation; LDL is the chief source in chronic stimulation.

  d **True.** Cytochrome P450 forms pregnenolone from cholesterol, the first common step in steroidogenesis.

  e **False.** Much of the production of steroid hormones takes place in the cytoplasm.

22 a **False.** Adrenal androgen production is probably regulated by auto-regulatory mechanisms within the adrenal cortex related to the centripetal flow of steroids from the outer layers. Adrenocorticotrophic hormone (ACTH) is ultimately critical.

  b **True.** An enzyme deficiency prevents the formation of cortisol from progesterone. Therefore there is a huge outpouring of ACTH in the adrenogenital syndrome.

  c **True.** Virilization occurs in the adrenogenital syndrome because of the diversion of cholesterol to androgen synthesis.

  d **False.** There is a block at P450, androgens are not formed, and so feminization is typical of congenital adrenal hyperplasia.

  e **False.** Only a small amount of adrenal oestrogen is made. Adrenal androgens are converted to oestrogen peripherally and thus replace ovarian oestrogen in postmenopausal women.

23 a **False.** The rate of release of adrenal corticosteroids corresponds to the rate of their synthesis.

  b **False.** Adrenal androgens are only important after peripheral conversion to testosterone and oestrogen. They contribute normally to axillary and pubic hair. Their production declines greatly after 10–15 years of age.

  c **True.** 90% of cortisol is bound to cortisol-binding globulin (CBG).

  d **False.** About 50% aldosterone is weakly bound to plasma proteins.

  e **False.** Half-life of cortisol is 60 min; of aldosterone, 20 min.

24 a **True.** Conjugated steroids are ultimately excreted either in the bile or (mostly) in the urine.

  b **True.** Firm cortisol binding to CBG gives it a long half-life.

c   **False.** In pregnancy oestrogen stimulates CBG synthesis.Thus total cortisol is raised but biologically active cortisol not elevated.

d   **True.** The cause is unclear.It may be due to the blunted response of brain cortisol receptors.

e   **True.** Cortisol-mediated feedback applies to all cortical hormones except aldosterone.

25   a   **True.** Cortisone and corticosterone are other glucocorticoids.

b   **False.** As for all steroid hormones, the basic action is to regulate gene expression.

c   **True.** Cortisol also increases gluconeogenesis, lipolysis, protein catabolism, and inhibits nucleic acid synthesis.

d   **False.** Cortisol is diabetogenic and inhibits glucose uptake.

e   **False.** Cortisol demineralizes bone. Fractures are common in Cushing's disease.

26   a   **True.** Thus cortisol delays repair of tissues.

b   **False.** Cortisol inhibits the mobilization of arachidonic acid and thus decreases the production of PGs and of eicosanoids.

c   **False.** Cortisol causes a decrease in circulating eosinophils and lymphocytes and a slight rise in neutrophils.

d   **True.** This is called the permissive effect of cortisol.

e   **True.** Usual cortisol levels promote normal function.

27   a   **True.** ACTH and cortisol show a typical negative feedback relationship.

b   **True.** Cortisol has a direct inhibitory effect on the release of both ACTH and CRH.

c   **False.** ACTH, and thus cortisol, appear in secretory bursts driven by body stresses (food, activity, environment, mood).

d   **False.** ACTH rhythm is probably independent of light/darkness. It is more related to the sleep/wake pattern.

e   **False.** ACTH rapidly increases cAMP; this leads to stimulation of cytochrome P450 side-chain cleavage enzyme. Thus ACTH (a glyco-protein) causes its final effects more rapidly than cortisol (a steroid).

28   a   **False.** A blood glucose level of < 2.2 mmol/l (40 mg%) is needed to raise the blood cortisol significantly.

b   **False.** Hypoglycaemia stimulates CRH.

c   **False.** This is because hypoglycaemia not only works via CRH but also activates ACTH by other means, e.g. via vasopressin.

d   **True.** Dexamethasone, 0.5 mg, 6 hourly for 2 days, suppresses ACTH.

e   **False.** Metyrapone inhibits cortisol production. It tests the capacity of the pituitary to respond to decreased circulating levels of cortisol.

29   a   **True.** ACTH is structurally similar to melanocyte stimulating hormone (MSH). ACTH is raised in Addison's disease.

b **False.** Classic symptoms of Addison's disease involve both gluco- and mineralocorticoids as well as sex steroids. Addison's disease is due to adrenocortical atrophy usually from autoimmune disease.

c **True.** The term Cushing's syndrome applies to any clinical state due to excessive prolonged glucocorticoid activity.

d **False.** Hyperpigmentation is likely from excess ACTH in secondary (pituitary) Cushing's syndrome.

e **False.** In adrenal tumours plasma cortisol is often elevated while the plasma ACTH is low or undetectable due to its suppression by high levels of cortisol.

30 a **True.** The 18-hydroxylase needed to make aldosterone is present only in the zona glomerulosa.

b **True.** Angiotensin II works via G proteins, inositol pathways, and raised cytoplasmic Ca^{2+} levels.

c **True.** The adrenal renin–angiotensin system results in major aldosterone effects via autocrine/paracrine pathways as well as via the classic pathway.

d **False.** ACTH is a weak stimulant of aldosterone. Its effect is short lived (< 24 h).

e **True.** Potassium probably depolarizes glomerulosa cells and allows Ca^{2+} entry through voltage-gated channels.

31 a **False.** In the kidney, aldosterone retains sodium in the distal renal tubule.

b **True.** Aldosterone reabsorbs Na^+ in the cortical collecting ducts to a much lesser extent than in the distal convoluted tubule.

c **False.** Aldosterone increases the concentration of the sodium pump in the basolateral membranes of mucosal and ductal target cells.

d **True.** Thus aldosterone acts on the ducts of salivary, pancreatic, and sweat glands.

e **True.** Aldosterone may increase the number of Na^+ and K^+ channels. It may also methylate the sodium ones.

32 a **True.** In salt-wasting congenital adrenal hyperplasia there is an almost complete lack of 21β-hydroxylase which is needed for both mineralo-corticoid and glucocorticoid synthesis.

b **True.** For example, autoimmune disease, HIV syndrome, and fulminant sepsis may all cause total adrenal cortical malfunction.

c **True.** This is the case if there is a reduced number or sensitivity of type 1 mineralocorticoid receptors (pseudohypoaldosteronism).

d **False.** Hyperkalaemia, hyponatremia, and hypotension are typical of adrenal insufficiency.

e **True.** It takes this time to synthesize the aldosterone-provoked proteins.

33 a **True.** Hyperaldosteronism and low plasma renin occurs usually in the 30–50 age group, and commonly there is a solitary adrenal adenoma.

b **True.** GSA is inherited as an autosomal dominant. Aldosterone formation extends to the zona fasciculata.

c **True.** Liquorice and carbenoxolone inhibit 11-hydroxysteroid dehydrogenase that normally inactivates cortisol and so prevent cortisol attaching to these receptors.

d **True.** Elevated plasma cortisol causes sodium retention and volume expansion.

e **False.** Aldosterone renal kinins renal PGs prorenin renin aldosterone (Note Bartter's syndrome). (= stimulates)

34 a **True.** There are insulin (beta), glucagon (alpha), SS (D-), and PP (F-) cells.

b **True.** A thin layer of α- and D-cells encircle large clumps of β-cells.

c **False.** In man pancreatic polypeptide probably inhibits the exocrine pancreas and relaxes the gall bladder.

d **False.** Pancreatic somatostatin causes local paracrine inhibition of the release of insulin and glucagon. It is also believed to generally 'pace' GIT functions.

e **False.** Pancreatic and hypothalamic SS are identical tetradecapeptides with a disulphide bond.

35 a **False.** Insulin contains two chains, A and B, with 21 and 30 amino acids respectively.

b **False.** Human insulin differs from pig insulin but in only one amino acid.

c **False.** Human insulin is complexed with zinc.

d **False.** Insulin is degraded significantly in both liver and kidney.

e **False.** The half-life of insulin is about 5 min.

36 a **True.** A liver-type glucose transporter is present in β-cells. Glucose favours uptake of calcium by β cell a prerequisite for insulin secetion.

b **False.** Amino acids, especially arginine and leucine, stimulate insulin release, as well as that of glucagon.

c **True.** Vagal stimulation increases insulin secretion, an effect that can be blocked by atropine.

d **False.** Beta agonists are hyperglycaemic via their effect on glucagon secretion although they may have a mild stimulatory effect directly on the beta cell in the pancreas. Alpha agonists appear to inhibit the beta cell directly.

e **False.** The response to oral glucose is greater due to GIP release.

37 a **False.** There is a low basal secretion of insulin.

b **False.** Insulin shows a biphasic response. The first phase is due to discharge of ready-made granules, the second is due to new insulin synthesis.

c **True.** IGFs are homologous to proinsulin. They have little intrinsic insulin activity (see growth hormone and IGF, Q48).

d **False.** Glucagon stimulates insulin release; insulin blocks glucagon release.

e **False.** Ketones are never more than minor stimulants of insulin release.

38 a **False.** There are separate receptors for insulin, IGF-I, and IGF-II.
   b **True.** Both insulin and its receptor consist of two polypeptide chains linked by S–S bonds. However, the receptor (mol. wt 150 000) is much larger than insulin.
   c **False.** Tyrosine kinase is on the inner domain of the insulin receptor. Insulin is bound externally.
   d **True.** This takes place by endocytosis. Most insulin receptors are recycled to the cell surface.
   e **False.** Insulin down-regulates its receptors in postnatal life. However, it up-regulates them in the fetus.

39 a **True.** Insulin is hypoglycaemic primarily by promoting an increased uptake of glucose by the tissues and an inhibition of hepatic gluco-neogenesis.
   b **True.** The antilipolytic action of insulin indirectly promotes further tissue uptake of glucose.
   c **False.** Insulin stimulates LPL. LPL hydrolyses chylomicrons and insulin promotes FFA uptake by cells.
   d **True.** Insulin counteracts the tonic effect of amines on hormone-sensitive lipase.
   e **False.** Amino acid transport is an independent effect of insulin. Insulin also stimulates protein biosynthesis.

40 a **False.** GLUTs allow facilitated diffusion (only down a concentration gradient).
   b **True.** This reflects the importance of glucose for all cells.
   c **True.** Very large amounts of GLUT-1 are found in the brain and red blood cells. It is also abundant in placenta, kidneys, and intestine.
   d **False.** GLUTs are bidirectional, saturable, Na^+ independent, and are not concentrative.
   e **True.** Insulin permits preformed GLUT-4 from intracellular organelles of myocytes and adipocytes to appear in the membrane by an exocytotic process.

41 a **False.** For example, excess ovarian androgens are formed in hyper-insulinaemic patients who none the less show severe resistance to the glucose-lowering effects of insulin (acanthosis nigricans).
   b **False.** Insulin antibodies are a rare cause of insulin resistance nowadays.
   c **False.** The linking of subcutaneous degradation of insulin and insulin resistance is widely disputed.
   d **True.** Both dominant and recessive patterns of inheritance can affect the integrity of the insulin receptor.
   e **True.** Defects in genes controlling GLUT-4, hexokinase, and glycogen synthase can cause insulin resistance.

42 a **True.** Antibodies to the insulin receptor are rare. Polyclonal IgG develops in some autoimmune diseases. A decrease in receptor number or affinity is more common.

b **True.** Cortisol, glucagon, catecholamines, and GH together and independently increase insulin resistance.

c **False.** Fewer and less sensitive insulin receptors occur in obese people.

d **True.** Resistance to insulin is increased in puberty, pregnancy, and old age.

e **False.** Resistance to insulin is increased in fever, sepsis, fasting, uraemia, and ketosis. Insulin secretion is also decreased in these states.

43 a **True.** Insulin stimulates glycogenesis and creates a glucose sump in liver cells.

b **False.** Neurons have an abundance of glucose transporter type 1 (GLUT-1) in the plasma membrane.

c **False.** Red blood cells have an abundance of GLUT-1.

d **False.** GIT and renal tubules also have Na^+-driven cotransport.

e **True.** Hypoinsulinaemia may cause sudden snowflake-like deposits throughout the cortex of the lens.

44 a **False.** Glucagon is a single chain of 29 amino acids.

b **True.** Glucose stimulates the release of somatostatin and in turn is inhibited by somatostatin.

c **True.** Glucagon also stimulates glycogenolysis in the liver.

d **True.** Glucagon is needed to prevent hypoglycaemia despite a decrease in insulin secretion at this time.

e **True.** Glucagon is lipolytic. FFAs and triglycerides thus liberated go to the liver to form glucose.

Anterior pituitary hormones and the endocrine hypothalamus (Qs 45–56)

45 *Concerning growth factors:*

a Growth hormone (GH) acts directly to increase cell growth.

b Insulin-like growth factors have structural homology to proinsulin.

c Vascular endothelial growth factor (VEGF) is stimulated by hypoxia.

d Endothelium is the richest source of platelet-derived growth factor (PDGF).

e Platelets are the main source of growth inhibitors (chalones).

46 *Growth hormone (GH):*

a is formed and secreted by the basophil cells of the anterior pituitary.

b from other primates is ineffective in humans.

c from human cadavers is the main source of GH in clinical practice.

d releases somatomedins (IGFs) from liver stores.
e its diabetogenic effect results from a direct action on target cells.

47 The secretion and release of growth hormone:

a is promoted by hypothalamic GH releasing hormone (GHRH) and is pulsatile in nature.
b is strongly inhibited by hypothalamic somatostatin (SS).
c is increased by exercise and stress.
d is promoted by DOPA and noradrenaline.
e is promoted by acetylcholine in the hypothalamus.

48 Regarding growth hormone and somatomedins (IGFs):

a More than 95% of IGF is tightly bound to plasma proteins.
b IGF-I remains fairly constant throughout postnatal life.
c Laron dwarfs have low GH and low IGF levels.
d African pygmies show a low level of both IGF-I and IGF-II.
e In hypothyroidism the release of GH in response to hypoglycaemia to arginine is reduced.

49 In the regulation of growth hormone:

a In plasma, GH exerts a negative feedback on the pituitary.
b 70% of total daily secretion of GH occurs during sleep.
c Acute doses of steroids reduce the plasma level of GH.
d The effect of cortisol on GH secretion is markedly abnormal in depression.
e Chronic fatigue syndrome (CFS, post-viral fatigue) is associated with a disturbance of GH regulation

50 Prolactin (PRL):

a is synthesized in the same cells as GH.
b plasma levels are only slightly higher in females than in males.
c inhibiting factor (PIF) is tonically released from the hypothalamus.
d release is stimulated by cortisol.
e excess in hyperprolactinaemia is treated with 5-HT or 5-HT agonists.

51 Prolactin:

a The most potent stimulus to PRL secretion is pregnancy.
b In pregnancy PRL levels rise steadily but fall just before term.
c PRL is the most frequent hormone produced in excess by pituitary tumours.
d PRL falls during sleep.
e PRL inhibits LH and FSH.

52 Adrenocorticotrophic hormone (ACTH), corticotrophin:

a is the smallest peptide hormone of the anterior pituitary.
b plays a role in the normal activity of skin melanocytes.
c acts as a growth factor on the adrenal cortex.

d with β endorphin and lipotropin are released following stimulation by corticotrophin releasing hormone (CRH).

e increases in Cushing's disease (adenoma of pituitary corticotrophs).

53 Controlling the release of adrenocorticotrophic hormone (ACTH):

a Corticotrophin releasing hormone (CRH) is a potent ACTH secretagogue.

b Arginine vasopressin (AVP) is equipotent with CRH as a secretagogue for ACTH.

c AVP is released along with CRH into the hypophyseal portal circulation.

d Angiotensin II stimulates the release of ACTH.

e Cortisol is the sole endogenous inhibitor of ACTH release.

54 Concerning the glycoprotein hormones of the anterior pituitary:

a They consist of thyroid stimulating hormone (TSH), luteinizing hormone (LH), and follicle-stimulating hormone (FSH).

b LH, FSH, and TSH are all made in different cells.

c Synthesis of TSH is stimulated by dopamine.

d External cold is a strong stimulant for TSH secretion.

e LH stimulates testosterone synthesis by the Leydig cells of testis.

55 The endocrine functions of the hypothalamus:

a All recognized neurohormones (NHs) from the hypothalamus are peptides.

b Neurosecretory cells send fibres to the median eminence.

c Many neurosecretory fibres go directly to the anterior pituitary.

d Thyrotrophin releasing hormone (TRH) is a large peptide.

e Interplay of somatostatin and GHRH causes a pulsatile GH secretion.

56 The endocrine hypothalamus:

a Neurohormones from the supraoptic and paraventricular nuclei travel in the hypophyseal portal system to the posterior pituitary.

b Interleukins such as IL-1, -2, -6, and TNF inhibit CRH release.

c Interleukins such as IL-1, -2, -6, and TNF inhibit AVP.

d AVP stimulates the release of CRH.

e Acting centrally, CRH activates the sympathoadrenal system.

Anterior pituitary hormones and the endocrine hypothalamus (Answers)

45 a **False.** Growth hormone acts via somatomedins (IGF-I, IGF-II) produced in the liver. It is uncertain if GH has any anabolic effects independent of IGFs. Note, however, that lipolysis and insulin antagonism are probably mediated directly by GH.

b **True.** Only high levels of IGFs stimulate the insulin receptor.

c **True.** VEGF is implicated in postnatal neovascularization of the retina in hypoxic conditions.

d **False.** PDGF is found mainly in platelets and also in the endothelium, neuroglia, and in fibroblasts. PDGF stimulates growth of smooth muscle and fibroblasts.

e **True.** For example, transforming growth factor B limits growth in most tissues.

46 a **False.** GH is made and secreted by acidophil cells (somatotrophs) of the anterior pituitary.

b **False.** Human GH has high degree of sequence homology with other primate GH. Note deaths from Creutzfeld–Jakob disease after injections of human GH.

c **False.** Bacteria in which the human gene has been expressed provide the source of GH in clinical practice since 1980.

d **False.** IGFs are not stored. GH stimulates their synthesis and release.

e **True.** GH stimulates lipolysis to FFA and glycerol and promotes glycogenolysis. It provides substrates for IGF's anabolic actions.

47 a **True.** GHRH is a powerful releasing factor. GH is released in a pulsatile manner.

b **True.** The hypothalamus is the most concentrated source of SS outside the GIT.

c **True.** Exercise, stress, fasting, anaesthesia, sleep, cold, and a low blood glucose all stimulate the release of GH by inhibiting SS and/or by stimulating GHRH.

d **True.** DOPA and NA stimulate GH release by releasing GHRH.

e **True.** ACH, via a muscarinic effect, inhibits somatostatin which then inhibits GH release.

48 a **True.** Most IGF is bound to IGF-binding protein 3. Thus the free plasma IGF is sufficient to promote growth but not insulin-like activity.

b **False.** IGF-I peaks at about 17 years and then declines rapidly. IGF-II remains fairly constant throughout postnatal life.

c **False.** Laron dwarfs have high GH, low IGF-I. The defective GH receptor on liver cells is inherited as an autosomal recessive trait.

d **False.** Only IGF-I is decreased. IGF-II is unable to promote normal growth on its own.

e **True.** This is because T_3 enhances GH production in the pituitary and GH actions on target cells.

49 a **False.** IGF-I (somatomedin C) has a negative feedback on the pituitary. High levels may also inhibit GH secretion by increasing somatostatin tone on somatotrophs.

b **True.** GH secretion is highest during sleep, especially during NREM sleep.

c **False.** Steroids inhibit somatostatin tone on somatotrophs and increase plasma GH.

d   **True.** The blunted cortisol response in depression increases somatostatin tone on the somatotrophs.

e   **True.** The symptoms of CFS resemble those of mild Addison's disease. It may be that CFS is a central form of Addison's with dysfunction of hypothalamic cortisol receptors (type 2 glucocorticoid receptors).

50   a   **False.** PRL is made in acidophilic lactotrophs. These comprise 30% of anterior pituitary cells.

   b   **True.** Plasma levels of PRC are about 5 ng/ml in males; 8 ng/ml in females.

   c   **True.** PRL facilitates the release of PIF (dopamine) from the hypothalamus. Thyrotropin releasing factor releases PRL.

   d   **False.** Cortisol inhibits PRL by blocking the serotoninergic release of PRL.

   e   **False.** Hyperprolactinaemia is treated with DOPA or DOPA agonists (bromocriptine, cabergoline).

51   a   **False.** The most potent stimulus for PRL secretion is the physical act of suckling.

   b   **False.** PRL rises steadily in pregnancy and is maximal at term.

   c   **True.** DOPA antagonists, 5-HT, phenothiazines, catecholamine depletors, and opiates also increase PRL secretion.

   d   **False.** PRL is maximal in sleep. It peaks 5–8 h after onset of sleep

   e   **True.** PRL helps maintain amenorrhea after parturition. Its main action is lactogenic on the breast.

52   a   **True.** ACTH is a single linear chain of 39 amino acids.

   b   **False.** The first 13 amino acids of ACTH are similar to those of melanotrophin. ACTH is only melanotrophic if present in excess.

   c   **True.** ACTH excess causes adrenal hyperplasia, a deficiency causes adrenal atropy.

   d   **True.** This is because these are integral products of ACTH biosynthesis.

   e   **False.** Hypersecretion of ACTH causes bilateral adrenal hyperplasia. Hypercortisolaemia suppresses CRH and ACTH secretion (from normal corticotrophs).

53   a   **True.** In humans CRH is by far the most important secretagogue of ACTH. This is not necessarily so in all species.

   b   **False.** AVP is not as potent a secretagogue for ACTH as for CRH. Note that both AVP and CRH are found in the paraventricular nucleus of the hypothalamus.

   c   **True.** Both CRH and AVP reach the corticotrophs via the local portal system.

   d   **True.** Angiotensin II is a weak secretagogue for ACTH. Other weak secretagogues include adrenaline, noradrenaline, and oxytocin.

   e   **False.** A potent hypothalamic corticotropin-release inhibitory factor is believed to exist. Other inhibitors include dopamine, somatostatin, and hypothalamic ANF. Inhibition is normally probably dominant over stimulation.

54 a **True.** All the glycoprotein hormones are made in the basophil cells of the anterior pituitary.

    b **False.** LH and FSH are made in same cell; TSH in special thyrotrophs.

    c **False.** Dopamine and somatostatin inhibit TSH release. TRH is chief stimulant for PIF (dopamine) secretion.

    d **False.** There is only a minor TSH response to external cold in adult humans.

    e **True.** Note that FSH induces LH receptors on Leydig cells.

55 a **False.** All hypothalamic neurohormones (NHs) are peptides except PIF (dopamine).

    b **True.** Neurohormones are released and concentrated in the median eminence.

    c **False.** NHs diffuse into capillaries in the median eminence and go to anterior pituitary via the portal system.

    d **False.** TRH is glutamic acid–histidine–proline in equimolar ratios.

    e **True.** Pulsatile release is also typical of gonadotrophin releasing hormone (GnRH) as well as of GH.

56 a **False.** Oxytocin and AVP travel via the hypothalamo-hypophyseal tract to the posterior pituitary.

    b **False.** Interleukins stimulate the release of CRH.

    c **False.** Interleukins stimulate the release of AVP.

    d **True.** AVP and CRH are formed very close together and are both released in stress.

    e **True.** CRH is central to the sympathoadrenal/stress reaction. Note that angiotensin II enters the subfornical region of the hypothalamus (lacks blood–brain barrier) and probably exerts an effect on central vasoconstrictor control.

Posterior pituitary hormones (Qs 57–59)

57 *Posterior pituitary hormones:*

    a Oxytocin and arginine vasopressin (AVP) are formed in the pars nervosa of the pituitary (posterior pituitary).

    b Oxytocin and AVP are stored with specific proteins in the pituitary.

    c Oxytocin effects are inhibited by progesterone.

    d Oxytocin secretion is stimulated by the enlarging uterus in pregnancy.

    e Oxytocin is a strong lactogenic hormone.

58 About arginine vasopressin AVP (antidiuretic hormone ADH):

a   A rise in plasma osmolality inhibits AVP release.

b   The cells that produce AVP are also osmoreceptors.

c   AVP attaches to high-affinity V_2 receptors in the epithelium of the distal nephron where even low doses open water channels (aquaporins).

d   Alcohol and /or exercise inhibit AVP release.

e   Increased plasma angiotensin II inhibits AVP release.

59 AVP (ADH):

a   AVP is a pentapeptide.

b   A sudden loss of as little as 50 ml of blood causes a volume-induced release of AVP.

c   Pressure receptors for AVP release are located in the aortic arch and carotid sinus.

d   Vasoconstrictor effects are seen even at low levels of AVP.

e   At physiological levels AVP stimulates the release of TSH.

Posterior pituitary hormones (Answers)

57   a   **False.** Oxytocin and AVP are are nonapeptides formed in the supraoptic and paraventricular nuclei of the hypothalamus.

b   **True.** Oxytocin and AVP are each attached to a specific neurophysin.

c   **True.** Progesterone antagonizes and oestrogen increases oxytocin effects.

d   **False.** Stimulation of areola and nipple are potent exciters of oxytocin release.

e   **False.** Oxytocin contracts the myoepithelial cells of breast; and contracts the uterus especially in late pregnancy.

58   a   **False.** Maximal AVP secretion occurs at around 295 mosm/kg. A rise or fall of 1% in osmolality evokes an appropriate AVP response.

b   **False.** Osmoreceptors are close by in the medial preoptic nucleus of the hypothalamus.

c   **True.** AVP binds to V_2 receptors which trigger cAMP-dependent insertion of intracytoplasmic aquaporin-2 vesicles into luminal wall ('membrane shuttle'; local PGs inhibit AVP here).

d   **True.** AVP release is inhibited by alcohol. It is also inhibited by exercise, alpha- and beta-adrenergic agonists, glucocorticoids, stress, pain, and hypoglycaemia.

e   **False.** Increased plasma angiotensin II stimulates AVP release.

59   a   **False.** AVP and oxytocin are both nonapeptides.

b   **False.** 5–10% of the blood volume must be lost before AVP rises.

c   **True.** A decrease in circulating volume leads to an increase in the secretion of AVP from the neurohypophysis via baroreceptors in aortic arch and carotid sinus, as well as via presure–volume detectors in atria and great veins.

d   **False.** Physiological levels of AVP are not vasoconstrictor.

e   **True.** AVP acts alone and synergistically with CRH in promoting TSH release. Note that AVP is present in high concentration in the portal blood of the anterior pituitary.

9

Reproduction and fetus

Male reproduction (Qs 1–5)

1 *In the testis:*

a Sertoli cells produce spermatozoa.
b Sertoli cells produce inhibin.
c tubular fluid is mainly produced by the Sertoli cells.
d total time of spermatogenesis is about 2 months.
e the blood–testis barrier maintains testis temperature around 34°C.

2 *Concerning male sex hormones:*

a Testosterone is the major steroid hormone produced by Leydig cells.
b Testosterone circulates unbound in the plasma prior to reaching its target cells.
c Testosterone is more potent than dihydrotestosterone.
d Leydig cells are quite sensitive to excessive heat.
e Castration after puberty in the male usually lessens sexual desire.

3 *Testosterone:*

a in females, mainly comes from the ovaries.
b is usually increased in plasma in acne.
c is regulated by LH which binds only to Leydig cells.
d inhibits release of LH and GnRH in the adult male.
e is essential for normal spermatogenesis.

4 *In male reproduction:*

a Only fully formed and motile spermatozoa are normally released into the lumen of the seminiferous tubules.
b Mean volume of the ejaculate should be about 5 ml.
c Injury to the lumbar sympathetic may cause retrograde ejaculation.
d A varicocele (dilated pampiniform venous plexus) has little or no relevance to fertility.
e Semen fructose levels are an index of seminal vesicle function.

5 *Semen:*

a Normal semen is alkaline.
b Sperms succumb rapidly to increased temperature.
c Prostate contributes about 1% of volume to ejaculate.
d Semen is stored in the seminal vesicles.
e Ageing sperms are associated with a high rate of chromosomal errors when they fertilize ageing oocytes.

Male reproduction (Answers)

1 a **False.** Sertoli cells provide physical and paracrine support for adjacent germ cells.

b **True.** Inhibin inhibits the release of pituitary FSH.

c **True.** Fluid in the seminiferous tubules arises from the Sertoli cells, bathes germ cells, and facilitates spermatogenesis and sperm survival.

d **True.** Spermatogonium ⇒ spermatocyte ⇒ spermatids ⇒ permatozoa.

e **False.** Scrotal location keeps testes cool. The barrier prevents exposure of germ cells to harmful circulating substances and the immune system, which would see the haploid gametes as foreign antigens.

2 a **True.** As for all steroid hormones, cholesterol is a common precursor of testosterone.

b **False.** About 95% of testosterone is bound (40% to albumin, 55% to globulins).

c **False.** Testosterone is converted to the powerful dihydro form in target cells.

d **False.** Leydig cells are better survivors than germinal epithelium in warm conditions, and testosterone secretion may continue despite damage to sperm production.

e **False.** After castration sexual desire and penile erection are usually normal. Some atrophy of prostate and seminal vesicles occurs.

3 a **True.** In females testosterone is made from progesterone via androstenedione. Some is also made in the skin and adrenal cortex.

b **False.** Although more testosterone may be biologically active due to a decrease in bound androgen in acne.

c **True.** LH binds to Leydig cells while FSH is specific for Sertoli cells. None the less FSH increases LH receptors on Leydig cells.

d **True.** Testosterone inhibits the release of LH and especially of GnRH (at hypothalamic level).

e **True.** Germ cells are probably direct target cells for androgens.

4 a **False.** Full motility and capacity to fertilize is achieved during the period in the epididymis (1–2 weeks).

b **False.** The mean volume of ejaculate is about 3.8 ml.

c **True.** Ejaculate goes into bladder because of failure of the bladder neck to close.

d **False.** Pregnancy rates of up to 40% can be expected after repair of a varicocele.

e **True.** 60% of semen comes from seminal vesicles (fructose 1.5–6.5 mg%).

5 a **True.** Semen has a pH of 7.35–7.50. It is rich in phosphate and bicarbonate buffers. Acid kills sperm.

b **True.** However, sperms can survive well at sub-zero temperatures (Note sperm banks).

c **False.** Semen is 20% prostate fluid, and is rich in Zn, acid phosphatase, and lipids.

d **False.** During ejaculation the seminal vesicles contract and wash the sperm-rich moiety onwards.

e **False.** Few if any deleterious effects arise from the ageing sperm or the few ageing oocytes that become fertilized.

Female reproduction (Qs 6–11)

6 *Regarding the ovary and oestrogens:*

a There are about 100 million follicles in the ovaries at puberty.
b Oestriol and oestrone are the chief hormones secreted by the Graafian follicle.
c Oestriol is the most important oestrogen in non-pregnant women.
d Plasma oestrogen levels increase in chronic liver failure.
e Oestrogens are critical for initiating the endometrial secretory phase of the menstrual cycle.

7 *In the neuroendocrine control of ovaries and uterus:*

a Very low levels of oestradiol cause a fall in FSH and LH.
b Rising levels of oestradiol cause an increase in FSH and LH.
c Inhibin selectively suppresses FSH secretion.
d Follicle rupture occurs at the same time as the LH and FSH surge.
e Only a single ovum is normally released during ovulation.

8 *Neuroendocrine control of the menstrual cycle:*

a FSH induces LH receptor acquisition on granulosa cells:
b Corpus luteum secretes little if any oestrogen.
c Progesterone lowers the body set temperature by 0.5°C.
d Pulsatile GnRH secretion is essential for normal ovarian function.
e The GnRH response is greatly influenced by local oestrogen.

9 *Gonadotropin releasing hormone (GnRH):*

a There are two forms, one for LH the other for FSH.
b Oestrogen reduces frequency of GnRH pulses.
c Clomiphene citrate increases the frequency of GnRH pulses.
d Thyrotrophin-releasing hormone (THR) stimulates GnRH release.
e Progesterone increases the frequency of GnRH pulses.

10 *Progesterone:*

a plasma levels in male and female are very similar.
b promotes secretory changes in uterus, Fallopian tube, and breast.
c increases uterine contractility in the non-gravid uterus.
d synthetic antagonists induce labour near term.
e in general, is anti-oestrogen on the myometrium.

11 *In selected dysfunctions of menstrual cycle:*

a Oligo- and amenorrhea occur in less than 20% of women athletes.
b These are likely to be due to the reduced fat-lean ratio in athletes.

c   Athletes are likely to have anovulatory cycles even in the absence of overt menstrual disturbance.

d   Primary amenorrhea is due to endometrial rather than ovarian or pituitary dysfunction.

e   Polycystic ovarian disease is associated with excess oestrogen.

Female reproduction (Answers)

6   a   **False.** There are about 400 000 follicles in the ovary at puberty. This is a decline from 1–2 million at birth.

  b   **False.** Oestradiol is the major oestrogen secreted by the follicles.

  c   **False.** Oestradiol is by far the most important oestrogen.

  d   **True.** This is because the liver normally degrades oestrogens. Note that feminization occurs in males with chronic liver failure.

  e   **False.** Oestrogens are critical for the proliferative phase of the menstrual cycle.

7   a   **False.** Low levels of oestradiol cause a sustained increase in FSH and LH, e.g. after castration.

  b   **True.** Rising levels of oestradiol cause a positive feedback. This is typical of the preovulatory surge of FSH and LH (oestrogen-enhanced GnRH receptor number).

  c   **True.** Inhibin is a peptide hormone from the granulosa cells (and Sertoli cells in the male). It is stimulated by the action of FSH on these cells.

  d   **False.** Hormone surge precedes rupture by 24–36 h. LH raises progesterone, PGs, and proteolytic enzyme levels to favour rupture.

  e   **True.** Only one follicle survives decreasing FSH support after first 7 days of cycle.

8   a   **True.** LH can exert ovulatory and follicle rupture effects because of the FSH-induced LH receptors on granulosa cells.

  b   **False.** The corpus luteum secretes both oestrogen and progesterone.

  c   **False.** Progesterone raises body temperature in the luteal phase.

  d   **True.** GnRH pulses occur 1–2 hourly in follicular and 4 hourly in the luteal phase of the menstrual cycle.

  e   **True.** GnRH response is influenced by local oestrogen and also by local neurotransmitters and opioids.

9   a   **False.** There is probably only one GnRH that controls both LH and FSH. Function of GnRH associated peptide is unclear.

  b   **True.** Exogenous oestrogen can be expected to produce anovulation.

  c   **True.** Clomiphene is a non-steroidal anti-oestrogen that stimulates ovulation. It is used to treat infertility.

  d   **True.** TRH stimulates the release of GnRH, possibly by a direct action or possibly by inhibiting local opioids.

e   **False.** Progesterone probably inhibits frequency of GnRH pulses. Progesterone inhibits LH release and therefore inhibits ovulation ('mini pill').

10   a   **False.** Very little progesterone is found in males. Its function is unknown in males.

b   **True.** Progesterone promotes secretory changes provided the target cells are primed by oestrogen.

c   **False.** Progesterone decreases contractility in both gravid and non-gravid uteri.

d   **False.** Synthetic progesterone antagonists are effective as early abortifacients, e.g RU 486.

e   **True.** Anti-oestrogen effects of progesterone include myometrial hyper-polarization, resistance to oxytocin, decreased number of oestrogen receptors.

11   a   **False.** Amenorrhea is as high as 40% in women who take vigorous exercise.

b   **False.** It is more likely that endorphins generated by exercise inhibit pulsatile GnRH secretion.

c   **True.** Athletes have anovulatory cycles because of markedly decreased spontaneous LH frequencies.

d   **False.** Primary amenorrhea refers to failure of the menses to be initiated at puberty. Secondary amenorrhoea occurs after a woman has had normal menstrual cycles, and pregnancy is the most common cause.

e   **False.** In polycystic ovarian disease the follicles fail to develop. Thus there is excess androgen and reduced oestrogen secretion. Women afflicted are very often hirsute as a result.

Pregnancy, the fetus, parturition, and the neonate (Qs 12–26)

12 *Pregnancy:*

a   Implantation begins with the attachment of the blastocyst to the endometrium.

b   The endometrial cells of the uterus form the trophoblast.

c   The trophoblast is destined to form the future placenta.

d   Both oestrogen and progesterone are necessary for the maintenance of pregnancy.

e   Most tests for pregnancy are based on the presence of excess oestradiol in the urine.

13 *Human chorionic gonadotropin (hCG):*

a   is a steroid hormone.

b   has actions very similar to LH.

c   appears at detectable levels in pregnancy in maternal serum and urine 10–15 days after the last missed period.

d   slide test for pregnancy is virtually 100% accurate when the subject is 3 weeks pregnant.

e   is only produced by the trophoblast.

14 Regarding placental hormonal function during pregnancy:

a   Relaxin is a hormone that comes from the corpus luteum.

b   Human placental lactogen (HPL) has an anti-insulin effect.

c   Prolactin (PRL) is formed in significant amounts by the placenta.

d   Placental progesterone decreases greatly towards end of pregnancy.

e   Removal of the corpus luteum after first 2 months of pregnancy is associated with premature labour.

15 Steroids in pregnancy:

a   Oestradiol is the chief oestrogen of pregnancy.

b   The placenta makes oestrogen by *de novo* synthesis.

c   Placental oestrogen promotes placental and uterine growth.

d   Measuring maternal oestriol can be used as a measure of fetal well being.

e   The fetal zone (FZ) of the fetal adrenal cortex exports steroids to the placenta.

16 In the placenta:

a   Umbilical arteries carry oxygenated blood from placenta to fetus.

b   Passage of nutrients across the placenta is easier as term approaches.

c   Maternal insulin readily passes the placental membrane.

d   Rh antibodies cross the placenta most readily in early pregnancy.

e   There is a block to the passage of ketones from mother to child.

17 Regarding fetal homeostasis:

a   Fetal glucose levels are lower than maternal ones.

b   Fetal hypoglycaemia is very likely with maternal fasting.

c   The PO_2 in the umbilical vein blood is about the same as the PO_2 in the uterine artery.

d   Amniotic fluid is actively secreted in the late stages of pregnancy.

e   Removal of excess amniotic fluid is mainly through the lining membranes of the amniotic sac.

18 About pregnancy:

a   Fetal temperature is normally about 1 °C above the mother's.

b   Strenuous exercise by mother can cause vasoconstriction in uterus.

c   Pregnancy is more hazardous to the health of older mothers.

d   Increased resistance to angiotensin 11 occurs in normal pregnancy.

e   Women who have had a completed pregnancy in the past have higher levels of prolactin than those who were never pregnant.

19 *Parturition:*

a   Towards term the progesterone:oestrogen ratio increases.
b   Towards term there is a fall in fetal glucocorticoid production.
c   Oxytocin/oestrogen may trigger uterine contraction.
d   Beta-adrenergic drugs inhibit labour by stimulating sympathetic nerves.
e   The active sympathetic receptors in myometrium during labour are mainly $\beta2$ in type.

20 *Changes in maternal physiology include:*

a   a cardiac output that is highest at mid-pregnancy.
b   oestriol causing a dramatic fall in vascular resistance in the uterus.
c   renal blood flow hardly increasing at all in pregnancy.
d   red cell mass (RCM) increasing in pregnancy.
e   an increase in minute ventilation adjusted closely to increased O_2 demands.

21 *Alterations in maternal physiology:*

a   Circulating procoagulants rise dramatically in pregnancy.
b   Blood flow to the liver increases significantly.
c   Plasma albumin rises in pregnancy.
d   Serum alkaline phosphatase rises considerably.
e   There is an increased renal clearance of iodine.

22 *Pregnancy-induced changes in maternal physiology:*

a   Plasma urea of the mother increases.
b   Tachypnea is typical of a normal pregnancy.
c   Cardiac output falls throughout the second stage of labour.
d   A bilateral parasternal murmur is normal.
e   The mother's oxygen consumption per g body weight/min is greater than that of the fetus.

23 *Regarding physiology at birth:*

a   Prior to birth, in fetal life the ductus arteriosus is kept open by local PGs.
b   Closure of the ductus is accelerated by a fall in local PO_2.
c   Prostaglandins should be given in cases of aortic or pulmonary artery atresia.
d   The blood volume at birth is the same as that in the adult (70 ml/kg body weight).
e   Mechanical squeezing of baby in vaginal delivery has survival advantages.

24 *In the neonate:*

a   the heart is very distensible.
b   the cardiac output per kg/body weight is about the same as in an adult.
c   oxygen consumption per kg in neonate is twice that of an adult.
d   neonate and infant rely predominantly on thoracic breathing.
e   there is poor urinary concentrating power.

25 *Neonatal and infant physiology:*

a   At full term most of the infants HbF has been replaced by HbA.

b   Virtually all HbF should be gone by 6 months of age.

c   At birth the neuronal tissue of the brain is largely unmyelinated.

d   Hypoxia in the infant causes hypoventilation and bradycardia.

e   Cerebral vasculature of infant is insensitive to hyperoxia.

26 *In the production of immunoglobulins:*

a   Neonatal serum IgG levels are approximately the same as maternal levels.

b   Adult levels of IgG are reached around 1 year of age through active synthesis.

c   Adult levels of secretory IgA are reached around 1 year of age.

d   Adult levels of serum IgA are reached around puberty.

e   Efficient IgM responses are not present until 4–5 years of age.

Pregnancy, the fetus, parturition, and the neonate (Answers)

12   a   **True.** Implantation occurs 7–8 days after fertilization.

b   **False.** Outer cells of blastocyst form the trophoblast.

c   **True.** Placenta and nutrient membranes come from the trophoblast.

d   **True.** Oestrogen and progesterone come from the corpus luteum in the first 2 months and from the placenta thereafter.

e   **False.** Most pregnancy tests are based on the presence of hCG in the urine.

13   a   **False.** hCG is a glycoprotein of similar structure to FSH, LH, and TSH.

b   **True.** hCG prolongs the life of corpus luteum for the first 2 months of pregnancy at a time when LH levels are low.

c   **True.** In a pregnancy test: hCG-latex particles will agglutinate if mixed with anti-hCG. Urine or serum in pregnancy (contains hCG) inhibits agglutination because it mops up anti-hCG.

d   **False.** Cross-reactions with LH give some false positives. The radioimmune assay test for hCG is accurate and quantitative.

e   **False.** Small amounts of non-placental hCG are made in the pituitary in both sexes and in almost all human cancers.

14   a   **True.** The placenta takes over hormone production from the corpus luteum as pregnancy progresses. Relaxin softens ligaments, softens the cervix, and inhibits uterine contractility.

b   **True.** HPL is like GH in structure and function. This is as occurs in gestational diabetes.

c   **True.** The rise in PRL in amniotic fluid is of placental origin.

d   **True.** A fall in placental progesterone helps initiate parturition.

e   **False.** Removal of corpus luteum before 2 months causes endometrial disintegration. Later removal is without effect.

15 a **False.** Oestriol is the chief oestrogen of pregnancy. It comprises 90% of urinary oestrogen.

   b **False.** For the placental synthesis of oestriol, intermediates must be carried to it from the fetal adrenal cortex via the fetal circulation.

   c **True.** Placental oestrogen also increases uterine contractility.

   d **True.** Maternal oestriol reflects function in the fetal adrenal cortex.

   e **True.** The FZ picks up maternal cholesterol and re-exports other steroids to the placenta for final processing.

16 a **False.** Umbilical arteries carry deoxygenated fetal blood to the placenta. They are the equivalent of pulmonary arteries in the adult.

   b **True.** Nutrients pass much easier in late pregnancy as the chorion layer thins out.

   c **False.** Only substances of mol. wt < 1000 pass the placental barrier easily.

   d **False.** Anti-D and other undesirable agents (e.g. alcohol, nicotine) cross most readily in late pregnancy.

   e **False.** Ketones cross as readily as glucose or amino acids.

17 a **True.** In terms of glucose consumption the fetus can be likened to an auxiliary maternal brain.

   b **True.** However, all fetal tissue can metabolize ketones readily.

   c **False.** Umbilical vein PO_2 is only 4.5 kPa (35 mmHg) because of high myometrial O_2 extraction and long diffusing distance across chorion. The PO_2 in the umbilical artery is 2.7 kPa (20 mmHg), and in the uterine artery is 13.5 kPa (100 mmHg).

   d **False.** Fetal urine is the major source of amniotic fluid in the latter part of pregnancy.

   e **False.** A fetus swallows 200–1000 ml amniotic fluid /day in late pregnancy. This correlates well with fetal urinary output.

18 a **True.** The fetus is highly insulated by amniotic fluid.

   b **True.** Strenuous maternal exercise causes vasoconstriction in uterus and may cause damaging fetal hypoxia.

   c **False.** There is no evidence to substantiate a higher risk from pregnancy *per se* in older women. However, the rate of stillbirth is twice as high in older (> 35 years) than in younger women (20–25 years).

   d **True.** Pre-eclampsia is associated with increased sensitivity to angiotensin II.

   e **False.** PRL is likely to be higher in nullipara. PRL and GH are putative cocarcinogens for breast cancer.

19 a **False.** Towards term progesterone falls, prolactin and oestrogen increase.

   b **False.** Glucocorticoids increase markedly towards term and amniotic PGs also increase.

   c **True.** Oxytocin/oestrogen may trigger labour given a critical fall in progesterone and an appropriate stretch of the myometrium.

d **False.** Beta-adrenergic drugs delay labour on the basis of direct inhibition of the myometrium.

e **False.** $\beta2$-receptors relax the uterus. Note that $\beta2$-type receptors also relax bronchioles and peripheral arterioles.

20 a **True.** Cardiac output peaks at this time to 30–40% above the non-pregnant state.

b **True.** Also in early pregnancy the rise in plasma oestriol parallels the increase in uterine blood flow.

c **False.** RBF increases 35–40%. GFR increases too but glomerular pressure stays the same.

d **True.** Red cell mass rises 20%. Haematocrit falls because ECF increases 40%.

e **False.** Increased ventilation is far in excess of new O_2 demands.

21 a **True.** The rise in procoagulants in pregnancy is probably an effect of oestrogen on the liver. Fibrinogen levels increase 2–4 times.

b **False.** Hepatic blood flow remains unchanged. Thus the proportion of the cardiac output going to the liver falls by about 35%.

c **False.** Plasma albumin levels fall because of plasma volume expansion. Total amount of circulating albumin remains virtually unaltered.

d **True.** Alkaline phosphatase rises especially in late pregnancy. It is mainly of placental origin.

e **True.** Goitre is more likely in pregnancy due to the increased renal clearance of iodine.

22 a **False.** Plasma urea and creatinine decline due to the increased GFR and RBF in pregnancy.

b **False.** In pregnancy there is an increase in the tidal volume but normally not in the rate of ventilation.

c **False.** Valsalva of second stage causes cardiac output to fall. Note that the cardiac output may increase up to 30% during uterine contraction.

d **True.** Murmurs are often present over the course of the hypertrophied internal mammary arteries.

e **False.** Fetal oxygen consumption per g is higher than that of the mother.

23 a **True.** Locally produced PGE_2 relaxes the media of the ductus arteriosus.

b **False.** Closure of ductus arteriosus is stimulated by the high PaO_2 of postnatal life.

c **True.** In aortic and pulmonary atresia life depends on keeping the ductus arteriosus patent using PGE_1 or PGE_2.

d **False.** Neonatal blood volume is 90–100 ml/kg. An extra 75–100 ml can be transferred to the baby by delayed clamping of the cord.

e **True.** Mechanical squeezing is linked with the ability to produce stress hormones and the squeezing out of debris (e.g. secretions, aspirate) from the respiratory tract.

24 a **False.** Ventricles poorly distensible. Cardiac output depends mainly on heart rate in neonate.

   b **False.** Neonatal cardiac output is 30–50% greater than adults per kg body weight.

   c **True.** O_2 consumption is about 6 ml/kg/min in the neonate. This is accommodated by the high levels of Hb and cardiac output.

   d **False.** Neonates and infants rely on diaphragmatic breathing. Ribs are poorly ossified, very flexible and more horizontal. Intercostal muscles are immature.

   e **True.** The poor urinary concentrating power of the neonate increases the susceptibility to hyperosmolar dehydration for at least the first year of life.

25 a **False.** Virtually all the newborn's Hb is in the HbF form.

   b **True.** Virtually all HbF has disappeared by 6 months of age apart from some exceptional genetic disorders.

   c **True.** Myelination continues until adolescence.

   d **True.** This paradoxical response to hypoxia is reversed in childhood.

   e **False.** The infant's cerebral vasculature is about three times more sensitive to hyperoxia than the adult's.

26 a **False.** Neonatal levels of IgG are about 150% of the mother's. Active secretion of IgG occurs through the human placenta.

   b **False.** Adult levels of IgG are reached at about 5–6 years of age.

   c **False.** Adult levels of secretory IgA are reached around 2–3 months.

   d **True.** Note that 0.1–0.3% of adults lack detectable serum IgA.

   e **False.** IgM synthesis starts from 33 weeks' gestation onwards. Adult levels are achieved at about 1 year of age.

Female menopause (Qs 27–28)

27 Menopause:

   a is caused by a failure in the secretion of gonadotrophins.

   b is characterized by a decline in inhibin secretion.

   c shows a correlation between age at onset and age of the menarche.

   d shows a close synchrony between menopausal flushes and LH pulses.

   e reduces the incidence of breast cancer due to the decline in oestrogen levels.

28 Postmenopausal (senile, idiopathic) osteoporosis:

   a is characterized by a decrease in the mineral/organic ratio of bone.

   b is rising in direct proportion to the ageing population in Europe and the USA.

   c is as common in elderly men as in elderly women.

   d is little affected by dietary factors.

   e is little influenced by ethnic origin.

Menopause (Answers)

27 a **False.** The menopause is due to a failure of the ovary to respond to gonadotrophins.

   b **True.** The decline in inhibin at the menopause contributes to the inordinate rise in FSH.

   c **False.** No correlation exists. The age of menopause differs in different societies.

   d **True.** None the less LH secretion *per se* is not responsible for menopausal flushes.

   e **True.** Oestrogen-dependent breast cancer declines after the menopause, but remember that only 25–30% of human breast cancers are hormone dependent.

28 a **False.** Osteoporosis shows a decrease mass per unit volume, i.e. osteoporosis is normal bone but there is less of it. Osteomalacia shows a decreased mineral/organic ratio.

   b **False.** The frequency of osteoporosis is rising disproportionately in Europe and in the USA.

   c **False.** Probably 1 in 4 elderly women and 1 in 20 elderly men will have significant problems related to osteoporosis in Europe and the USA.

   d **False.** Risk factors for osteoporosis include: alcohol, protein, PO_4, NaCl, and a decrease calcium in the diet. Caffeine may also be a risk factor.

   e **False.** Caucasians/Orientals are most at risk from osteoporosis.

10

Nutrition and metabolism

Food and energy (Qs 1–3)

1 Regarding food and energy potential:

a   1 g of carbohydrate (CHO) yields about 4 kcal in the body.

b   1 g of fat has the same energy yield as 1 g of ethanol.

c   1 kilojoule (the SI unit of energy) = 10 kcalories.

d   100 g whole milk yields about 10 kcal.

e   An average hen's egg weighs about 100 g.

2 In metabolism:

a   The basal rate of metabolism is more closely related to height than to body area.

b   The basal rate is about 40 kcal/m^2/h.

c   Basal levels fall in starvation.

d   The respiratory quotient (RQ) of fat is greater than unity.

e   The basal rate of metabolism is greater in females than in males when account is taken of body weight/height/surface area.

3 Regarding energy usage and requirements:

a   In normal adults, basal metabolism remains roughly the same throughout life.

b   The basal metabolic rate (BMR) is much greater in a child than in an adult.

c   Metabolic energy needs/kg are roughly similar in child and adult.

d   The thermogenic effect of food (specific dynamic action) is most marked with fats.

e   The three major food groups (CHO, protein, fat) are interchangeable in the body.

Food and energy (Answers)

1   a   **True.** 1 g of CHO yields about 4 kcal when it is completely oxidized in the body.

b   **False.** 1 g fat = 9 kcal; 1 g ethanol = 7 kcal, and 1 g protein = 4 kcal.

c   **False.** 4.1868 kJ = 1.0 kcal.

d   **False.** 100 g of whole milk (cow's or human) provides 60–80 kcal.

e   **False.** The weight of an average hen's egg is about 50 g.

2   a   **False.** The Du Bois formula, used to calculate area, incorporates the weight and height of the subject.

b   **False.** This is the basal metabolic rate (BMR). Average adult requires 1.0–1.5 kcal/min to maintain basal metabolism.

c   **True.** Basal metabolism falls to about 20% in starvation. This is related to the reduction in lean body mass.

d **False.** RQ = CO_2 produced / O_2 consumed. RQ for fat = 0.7; RQ for CHO = 1.0.

e **False.** The basal metabolism is about 10% higher in males. It rises by about 15% in pregnancy.

3 a **False.** The basal metabolism falls with age in both sexes.

b **True.** The BMR falls from over 50 kcal/m²/h in childhood to the adult value by around 20 years of age.

c **False.** A child aged < 5 years requires 100–200 kcal/kg/day (three times the adult needs).

d **False.** The thermogenic effect of food occurs irrespective of the type of food eaten. It appears to be most marked with proteins.

e **False.** Neither fat nor CHO can make protein directly because they lack nitrogen; otherwise interchangeability applies.

Protein and related metabolism (Qs 4–8)

4 *Protein metabolism:*

a There are about 100 different amino acids important in human nutrition.

b Amino acids can be made in the body by aminating non-amino acids.

c In transamination there is a transfer of NH_3 to a keto acid.

d Ammonia is produced by oxidative deamination.

e Oxidative deamination only occurs in the liver.

5 *Protein requirements:*

a The minimum requirement in an adult is 1 g/kg body weight/day.

b The minimum requirement for a child under 5 years old is about 2.0 g/kg body weight/day.

c The biological value (BV) of a protein depends on its ability to form CHO and fat *in vivo*.

d The chemical score takes account of the essential amino acids in a protein.

e A negative nitrogen balance: nitrogen intake exceeds nitrogen output.

6 *In protein turnover:*

a DNA (deoxyribonucleic acid) is identical in every cell in the body (excluding ova and sperms).

b Proteins are synthesized in the nucleus of cells.

c Messenger RNA (mRNA) brings the appropriate amino acids together prior to protein synthesis.

d Most of the NH_3 formed by deamination is converted to urea.

e The formation of urea is critically dependent on the mitochondrion.

7 *In protein metabolism:*

a Oral methionine can provoke or unmask hyperhomocysteinaemia.

b Actin is the most abundant protein in mammalian cells.

   c Dying cells release actin into the ECF.
   d Glycine is the most common amino acid in the body.
   e Failure to make citrulline in the liver can be fatal.

8 *The urea cycle:*

   a The capacity to make urea diminishes in starvation.
   b The capacity to make urea diminishes in hypercatabolic states.
   c Glutamine formation is increased if NH_3 is not incorporated into the urea cycle.
   d It directly provides intermediaries for the citric acid cycle.
   e Urea contributes to the formation of muscle creatine.

Protein and related metabolism (Answers)

4 a **False.** There are about 20 amino acids important in human nutrition, of which 9–10 are essential (i.e. cannot be synthesized in the body).
   b **True.** For example, glutamic acid and alanine can be made from ketoglutarate and pyruvate provided other amino acids are present to supply the necessary amino groups.
   c **False.** Transamination involves transfer of a NH_2 group to a keto acid.
   d **True.** Amino acid + O_2 (of H_2O) becomes a keto acid + NH_3.
   e **False.** Oxidative deamination occurs in the liver and in the tissues.

5 a **False.** As little as 0.5 g of protein/kg/day is adequate provided it is of high biological quality.
   b **True.** The minimum protein requirement assumes a full range of essential amino acids and sufficient CHO or fat to spare protein for vital functions.
   c **False.** The BV relates to the % of N_2 retained from a given amount of protein N_2 that has been digested and absorbed.
   d **True.** Egg protein is the reference point, with a score of 100.
   e False A negative nitrogen balance occurs when N_2 in < N_2 out, e.g. in hypercatabolism.

6 a **True.** DNA is also unique for each individual.
   b **False.** Proteins are made in the ribosomes in the cytoplasm (cytosol).
   c **True.** mRNA presents the sequence of amino acids to the ribosome; transfer RNA collects them.
   d **True.** Urea is made from 2 NH_3 and 1 CO_2 in the urea cycle.
   e **True.** The first stage, ornithine citrulline, is ATP dependent and occurs in the mitochondrion.

7 a **True.** Methionine is the precursor of homocysteine.
   b **True.** Cell mobility and changes in size depend on the ability of actin to polymerize into actin filaments.

c   **True.** Actin is cleared from the plasma by gelsolin and a Gc protein. Intravascular free actin filaments are highly dangerous.

d   **False.** Glutamic acid is the most common amino acid in the body.

e   **True.** A recessive X-linked deficiency of carbamyl transferase (ornithine citrulline) can cause fatal hyperammonaemia.

8   a   **False.** All five enzymes of the urea cycle are up-regulated whenever there is an increase in the rate of hepatic amino acid catabolism.

b   **False.** As in 8a. Other protein metabolic and gluconeogenetic enzymes are also increased in hypercatabolic states.

c   **True.** Excess free ammonia leads to diminished ketoglutarate, an important substrate for the Krebs tricarboxylic cycle.

d   **True.** The urea cycle provides argininosuccinate, a common substrate for both the urea and the citric acid cycle.

e   **True.** Urea contributes to the formation of muscle creatine via hydroxyurea and hydroxylamine. The significance of this is not clear.

Carbohydrate and related metabolism (Qs 9–25)

9 Carbohydrate (CHO):

a   In a well-balanced diet carbohydrate should contribute no more than 50% to one's daily calorie needs.

b   In much of our diet, CHO comes in the form of glucose.

c   Cellulose is broken down to mono- and disaccharides before absorption.

d   A good diet should contain up to 10 g of polysaccharide fibre (cellulose, hemicellulose, pectins, gums, mucilages) per day.

e   A normal adult can only absorb about 120 g of glucose/h.

10 Usage of glucose:

a   50% of the glucose absorbed is stored as glycogen.

b   Only 10% of ingested glucose is normally converted to fat.

c   The rate of glucose phosphorylation in the hepatocyte varies with fluctuations in the level of plasma glucose.

d   Beta cells and hepatocytes both contain an insulin-independent glucose tansporter (GLUT).

e   Neural tissue consumes about the same amount of glucose as the total red cell mass.

11 In glucose metabolism:

a   Muscle cells lack glucose-β-phosphatase, the enzyme necessary to split glucose 6-PO_4 to free glucose and phosphate.

b   Glycolysis is the breakdown of glycogen to glucose.

c   Substrate phosphorylation involves the formation of ATP by transfer of a phosphate group to ADP from a metabolic intermediate.

d Oxidative phosphorylation occurs at a low level in glycolysis.

e About 40% of the energy released by the complete combustion of glucose to CO_2 and H_2O is trapped as ATP.

12 The formation and metabolism of pyruvate:

a Glycolysis is of little importance in the resting state in man.

b Pyruvate levels rise in beri beri.

c Pyruvate acetyl CoA is reversible if oxygen is unavailable.

d The carboxylation of pyruvate to oxaloacetate is an irreversible reaction.

e Alanine is a ready source of pyruvate.

13 In the citric acid (Krebs) cycle:

a Acetyl CoA is the only entry point for metabolites.

b Pyruvate is ultimately the only source of acetyl CoA.

c The vast majority of acetoacetyl CoA is formed in the liver.

d A falling level of ATP inhibits the cycle.

e Fluoroacetate directly inhibits key enzymes in the cycle.

14 Intermediary metabolism:

a In normal people, at rest and on a balanced diet there are no ketones in the plasma.

b The liver is the major site for destruction of ketone bodies.

c The liver is the major site for the breakdown of lactic acid.

d Acetoacetate is formed virtually only in the liver.

e If oxaloacetic is diverted to form glucose then ketosis is likely.

15 In intermediary metabolism:

a The Embden–Meyerhof (EM) pathway is the metabolic route for pentose metabolism.

b The hexose-monophosphate shunt (HMP shunt) is an alternative to the EM pathway.

c The Cori cycle refers to the production of lactate from pyruvate.

d Protein sparing means that at times other foodstuffs can take over certain metabolic activities normally carried out by proteins .

e Glucose counterregulation refers to the ability of the kidney to conserve glucose.

16 About glycogenolysis and gluconeogenesis:

a Glycogenolysis occurs by a reversal of the glycogenetic pathway.

b Gluconeogenesis is essential for normal living and normal activities.

c Gluconeogenesis is inhibited by insulin.

d Catecholamines and glucagon stimulate gluconeogenesis.

e Cortisol is a weak inhibitor of gluconeogenesis.

17 About gluconeogenesis:

a Gluconeogenesis is normally confined to the liver.

    b Cerebral gluconeogenesis makes a significant contribution in starvation.

    c All amino acids are substrates for gluconeogenesis.

    d Fatty acids are a normal substrate for gluconeogenesis.

    e Lactic acid is a substrate for gluconeogenesis.

18 *In hypoglycaemia:*

a clinical symptoms may occur when the blood glucose falls to about 3.3 mol/l (60 mg%) and is to be expected around 2.2 mmol/l (40 mg%) using arterialized venous blood. Note: 18 × mmol/l = mg/dl.

b symptoms appear at higher levels of blood glucose in the neonate than in the adult.

c neuroglycopenic symptoms are rapidly relieved by glucose.

d catecholamine-mediated symptoms are rapidly relieved by glucose.

e symptoms due to altered cerebral metabolism come on fairly slowly.

19 *Hypoglycaemia:*

a Postprandial biochemical hypoglycaemia occurs 2–3 h after a heavy CHO meal.

b Postprandial clincial hypoglycaemia is very liable to occur after gastric surgery.

c Biochemical hypoglycaemia is liable to occur in normal subjects after a 12 h complete fast.

d Exercise-induced hypoglycaemia is very common in athletes.

e Women withstand hypoglycaemia better than men.

20 *Alterations in blood glucose produced by:*

a Alcohol is more likely to provoke a hyper- than a hypoglycaemia.

b Leucine sensitivity can provoke a hypoglycaemia.

c Transitory hypoglycaemia in the newborn is related to hepatic immaturity.

d Hypoglycaemia of Jamaican vomiting sickness is due to sulphonylureas in the ackee fruit.

e Hypoglycaemia is a hazard of total parenteral nutrition (TPN).

21 *Regarding the handling of glucose:*

a In a normal glucose tolerance test (GTT) the blood glucose does not exceed 10 mmol/l (180 mg%) within 2 h.

b A properly performed GTT is reliable irrespective of the time of day at which it is done.

c Plasma glucose is normally the same as that of whole blood.

d Plasma glucose in fetus is usually higher than that of the mother.

e In postnatal life CSF glucose is always abnormal if it is less than 50% of the blood value.

22 *In diabetes mellitus (DM):*

a Antibodies to islet cells are found in less than 10% of cases of type 1 diabetes (insulin-dependent DM, IDDM).

b Elevated plasma insulin may occur in type 2 diabetes (non-IDDM).
c The non-insulin-dependent form can be found in young patients.
d Good control of gestational diabetes reduces both maternal and fetal mortality.
e Tight control of blood glucose reduces morbidity.

23 *In diabetes mellitus:*

a there is a decrease in blood fatty acids due to increased fat usage.
b a hypoaminoacidaemia occurs as part of deranged protein metabolism.
c plasma glucagon is lower than normal.
d the Somogyi phenomenon is the decreasing resistance to insulin sometimes found as diabetics get older.
e exogenous insulin is only of use if delivered by injection.

24 *In hyperosmolar diabetic ketoacidosis:*

a the arterial blood pH is usually less than 7.3.
b the plasma bicarbonate is usually between 20 and 25 mmol/l.
c the serum glucose is usually between 10 and 15 mmol/l.
d the calculated osmolality is greater than 330 mosmol/l.
e the patient is in severe K^+ depletion.

25 *With diabetes mellitus:*

a Hyperosmotic non-ketotic coma constitutes about 50% of diabetic decompensation episodes.
b Troglitazone (sensitizes insulin receptors) is useful in treating IDDM.
c The glycaemic index refers to the % of glucose in specific foods.
d Isocaloric helpings of different foods, e.g. rice and potatoes, produce the same postprandial rise in blood sugar.
e a diabetic 'coma' has a serum bicarbonate is less than 10 mmol/l.

Carbohydrate and related metabolism (Answers)

9 a **False.** Ideally CHO provides > 60%; protein about 10%, and fat < 30%.
b **False.** There is little glucose *per se* in diet. Most CHO is taken as starch or sucrose.
c **False.** Cellulose is an unbranched indigestible polysaccharide.
d **False.** A good diet should contain up to 100 g fibre/day. Non-polysaccharide lignins contribute a variable amount to this.
e **True.** Controlled delivery of chyme prevents inordinate hyperglycaemia after a meal rich in carbohydrates.

10 a **False.** Only 10% of ingested glucose is stored as glycogen.
b **False.** Around 40% of ingested glucose is normally laid down as fat.

    c **True.** Fluctuations in plasma glucose directly affect its rate of phosphorylation by hepatic glucokinase. This enzyme is also found in the beta cells of the pancreas.

    d **True.** GLUT-2 facilitates a very rapid equilibration of ECF and ICF glucose across the walls of beta cells and hepatocytes so that glucokinase becomes the limiting factor in glucose entry. GLUT-2 is also present in the walls of the small gut mucosa and the PCT of the kidney.

    e **False.** Glucose consumption by the brain is 120 g/day; RBCs consume 36 g/day.

11   a **True.** Muscle cannot directly contribute glucose to the blood.

    b **False.** Glycolysis is the breakdown of glucose to pyruvate.

    c **True.** Substrate phosphorylation can occur in the general cytoplasm and does not require a mitochondrion, e.g. in the red cell.

    d **False.** Oxidative phosphorylation only occurs in the mitochondrion and in the presence of oxygen (Krebs cycle).

    e **True.** About 40% of metabolic energy is trapped as ATP in aerobic metabolism. Only 2% is trapped as ATP in anaerobic states.

12   a **False.** Glycolysis is the major source of pyruvate and the only source of energy for the RBC and the crystalline lens. It is also vital for muscles contracting under anaerobic conditions.

    b **True.** Thiamine is necessary for the conversion of pyruvate to acetyl coenzyme A.

    c **False.** Pyruvate to acetyl CoA is irreversible and is oxygen-dependent.

    d **False.** Pyruvate to oxaloacetate is reversible and is substrate driven.

    e **True.** Alanine, a glucogenic amino acid, accesses the Krebs cycle via pyruvate.

13   a **False.** Substrate entry into the Krebs cycle is possible at any intermediate stage in the cycle.

    b **False.** Ketogenic amino acids and fatty acids also form acetyl CoA.

    c **False.** Most tissues combine two acetyl CoA molecules to form acetoacetyl CoA.

    d **False.** A rising level of ATP inhibits two key initial cycle reactions (pyruvate to acetyl CoA; acetyl CoA to citrate).

    e **False.** Fluoroacetate is changed to fluorocitrate in the body and this substance then inhibits aconitase, the enzyme needed for citrate to isocitrate.

14   a **False.** Ketonaemia is normally < 1 mg% .

    b **False.** Ketones are destroyed peripherally (in extrahepatic tissue).

    c **True.** Lactic acid is metabolized in the liver.

    d **True.** Acetoacetate, acetone, and betahydroxybutyrate (all ketones) are formed exclusively in the liver.

    e **True.** For example, during fasting the Krebs cycle runs down due to lack of oxaloacetate to convert to glucose via phosphoenolpyruvate. The extra acetyl CoA formed from fat is diverted to ketones which are important sources of metabolic energy in fasting (and in diabetics).

15 a **False.** The EM pathway is the common glycolytic chain to pyruvate via fructose and triose sugars.

b **True.** Glucose 6-PO_4 reaches triose sugars via glucuronic acid.

c **False.** The Cori cycle is the export of muscle lactate to the liver; the hepatic formation of glucose, and the subsequent export of this glucose to muscle.

d **True.** Protein sparing occurs when other foods take over non-essential activities mediated by proteins, e.g. energy production.

e **False.** Counter-regulation is the ability of specific hormones to counteract hypoglycaemia and restore blood glucose to normal.

16 a **False.** Two distinct pathways are involved in glycogenolysis and glycogenesis. Each one inhibits the other.

b **True.** The brain requires 120 g glucose/day. Only 20 g is readily available in body fluids.

c **True.** Inhibition of gluconeogenesis is part of the hypoglycaemic effect of insulin.

d **True.** Adrenaline and glucagon are the two most powerful 'counter-regulatory hormones'. Normally glucagon is the more important.

e **False.** Gluconeogenesis is moderately stimulated by cortisol and only weakly stimulated by growth hormone.

17 a **False.** 90% of gluconeogenesis takes place in the liver; 10% in the kidney.

b **False.** Renal gluconeogenesis accounts for 30% of the total in starvation.

c **False.** Only the glucogenic amino acids can form glucose.

d **False.** Only the glycerol portion of fat is glucogenic.

e **False.** Gluconeogenesis implies glucogenesis from a non-carbohydrate source

18 a **True.** Impaired brain function has occurred at 2.8 mmol/l (50 mg%) during acute insulin-induced hypoglycaemia in healthy adults. Note that biochemical hypoglycaemia does not imply the presence of symptoms.

b **False.** Symptoms of hypoglycaemia appear at 1.6 mmol/l (30 mg%) in full-term neonate or 1.1 mmol/l (20 mg%) in the premature baby.

c **False.** Neuroglycopenic symptoms often take several hours to reverse.

d **True.** Adrenosympathetic manifestations of hypoglycaemia include anxiety, trembling, tachycardia, palpitations, tremor, sweating, and light-headedness.

e **True.** Neuroglycopenic manifestations of hypoglycaemia include headache, mental dulling, confusion, amnesia, abnormal behaviour, dizziness, convulsions, and coma. They come on fairly slowly.

19 a **True.** An exaggerated insulin response is provoked by excessive carbohydrate loading. It will not produce clinical hypoglycaemia in a normal person.

b **False.** Hurried gastric emptying and rapid absorption of excess CHO may cause a biochemical hypoglycaemia (19a). There is no clear evidence that hypoglycaemia *per se* causes any post gastric surgery symptoms.

c **True.** A normal person may have a blood glucose well below 2.8 mmol/l (50 mg%) when fasting. The liver produces 200 g /day, enough for body needs, and leaves a comfortable excess for ordinary muscle activity.

d **True.** Once the exercise is stopped normoglycaemia rapidly returns.

e **True.** The relatively greater resistance of women to hypoglycaemia may be related to the greater capacity of women to produce ketones in the brain.

20 a **False.** Hepatic metabolism of alcohol to acetaldehyde to acetyl CoA involves the reduction of NAD^+ to NADH. NAD is necessary for gluconeogenesis.

b **True.** Leucine can provoke an exaggerated insulin response in some children.

c **True.** Neonatal hypoglycaemia may become prolonged and serious, especially if sepsis is present.

d **False.** 'Bush tea' from ackee fruit contains hypoglycin A, a powerful hypoglycaemic agent.

e **True.** Insulin-induced hypoglycaemia may occur in TPN where insulin is used to counteract the hyperglycaemia of hypertonic glucose infusions.

21 a **True.** In the GTT blood glucose should also be back to normal after 2 h.

b **False.** The GTT may mislead if done late evening or near dawn (when there is often an unexplained rise in blood glucose, probably due to an increase in cortisol).

c **False.** Plasma glucose is usually 15% higher than that in whole blood (RBCs contain little glucose).

d **False.** Fetal glucose values are usually 2/3 that of the mother (this reflects a greater dependence of the fetus on glucose as a metabolic fuel).

e **True.** However, a very low CSF glucose is often found in pre-term babies.

22 a **False.** Antibodies to the islet cells are found in 60–90% of newly diagnosed type 1 diabetics.

b **True.** NIDDM may show considerable insulin resistance.

c **True.** In many young patients with NIDDM the condition is inherited as an autosomal dominant trait.

d **True.** Tight control of blood glucose in DM has reduced the perinatal and maternal mortality dramatically.

e **True.** The balance of evidence now shows this to be the case.

23 a **False.** Hyperlipidaemia occurs in DM because of the lack of the antilipolytic effect of insulin.

b **False.** In DM there is a hyperaminoacidaemia because of increased protein breakdown and a failure to transport amino acids into cells.

c **False.** This may be due to the inability of glucose to enter alpha cells and turn off glucagon secretion in the absence of insulin.

d **False.** The Somogyi phenomenon refers to insulin resistance resulting from persistent hyperglycaemia due to an excess of counter-regulatory hormones.

e **False.** Intranasal and aerosol formulations of insulin hold great promise.

24 a **True.** The arterial pH usually falls below pH 7.3 in hyperosmolar diabetic ketoacidosis (HDK) due to the severe ketoacidosis.

  b **False.** Plasma bicarbonate is usually less than 20 mmol/l in HDK.

  c **False.** Serum glucose is usually greater than 15 mmol/l (270 mg%) in HDK.

  d **True.** Calculated osmolality is given by $2Na^+$ [mmol] + Glu [mmol] + BUN [mmol].

  e **True.** Patients with HDK usually have lost 1 mol of potassium in the urine. Note there is a deceptive normo- or hyperkalaemia due to dehydration.

25 a **False.** Only 5% of diabetic coma is non-ketotic.

  b **False.** Troglitazone is of use in about 50% of cases of NIDDM.

  c **False.** The glycaemic index refers to the ability of a food to produce postprandial hyperglycaemia.

  d **False.** The ensuing hyperglycaemia from isocaloric food helpings varies greatly and unexpectedly.

  e **True.** The term coma is used if HCO_3^- < 10 mmol/l irrespective of the state of consciousness.

Fat and related metabolism (Qs 26–35)

26 Fat:

  a Fat accounts for 10–20% of body weight in an average adult in a Western society.

  b In general, females contain more fat per kg than males.

  c In general, triglycerides are acidic in nature.

  d Sterols are solid alcohols.

  e The degree of saturation of a fatty acid increases as the number of its double bonds increases.

27 Omega-3 fatty acids(eicosapentanoic, EPA; docosahexanoic acid, DHA):

  a These are small fatty acids comprised of three 2-carbon groups.

  b They have their only double bond at the omega (terminal) position.

  c Linoleic acid is rapidly converted to EPA and DHA in the body.

  d Linoleic acid is found in abundance in fish-oils.

  e Omega-3 acids cause a hypolipidaemia.

28 Chylomicrons:

  a Chylomicrons go directly to the liver via portal venous blood.

  b About 90% of the weight of a chylomicron is triglyceride.

  c Hormone-sensitive lipase (HSL) removes FFA and glycerol from circulating chylomicrons.

    d Phospholipase is important in the degradation of chylomicrons.

    e The chylomicron remnant (30–80 nm) is small enough to pass through fenestrations in the liver sinusoids.

29 Blood lipoproteins:

    a Very low density lipoproteins (VLDLs) are the chief way in which triglycerides are transported from the liver to extrahepatic tissues.

    b VLDL is the immediate precursor of low density lipoprotein (LDL).

    c The liver removes almost all the LDL that is cleared in 24 h.

    d High density lipoprotein (HDL) is made in the liver.

    e HDL is involved in reverse cholesterol transport.

30 Cholesterol:

    a is mostly in the body as a constituent of cell membranes.

    b intake is necessary for normal health.

    c is transported in the blood with lipoproteins.

    d stimulates hydroxymethylglutaryl CoA (HMG-CoA) reductase.

    e stimulates the production of LDL receptors.

31 Cholesterol:

    a A rise in the number of LDL receptors causes a rise in serum cholesterol.

    b An increase in extrahepatic lipoprotein lipase activity causes a decrease in serum cholesterol.

    c Lovastatin lowers cholesterol by inhibiting HMG-CoA reductase.

    d The atherogenic nature of LDL is clearly established.

    e HDL is anti-atherogenic and LDL is atherogenic.

32 Fat:

    a Linoleic acid is the only essential fatty acid.

    b β-Oxidation of fatty acids only takes place in the mitochondria.

    c A typical adipocyte contains multiple droplets of fat in the cytoplasm.

    d Much of the depot fat originates from hepatic glucose.

    e The number of adipocytes is fixed for life at birth.

33 Functions of fat:

    a In terms of size of body fuel reserves: fat > CHO > protein.

    b Saturated fats have a lower freezing point than unsaturated ones.

    c Unsaturated fatty acids are placed more superficially and are more plentiful in the extremities.

    d Resistance to volatile anaesthetics is proportional to fat depot size.

    e Fat is a good source of endogenous water.

34 Alcohol (ethanol):

    a is chiefly absorbed in the jejunum.

    b increases lactic acid formation.

    c is generally metabolized more rapidly in women than in men.

d decreases sympathetic tone to vessels in skeletal muscles.
e from moderate drinking (1–3 drinks/day) is associated with increased serum concentrations of high density cholesterol.

35 *Effects of alcohol (ethanol) include:*

a decreased hepatic nicotinamide (NAD^+).
b increased metabolism of alcohol as blood levels rise.
c habitual drinkers metabolizing alcohol faster than non-drinkers.
d inhibition of sympathetic vasoconstrictor tone in skeletal muscle.
e increased cardiac output, chiefly by increasing the heart rate.

Fat and related metabolism (Answers)

26 a **True.** The amount of fat varies considerably between individuals.
b **True.** Fat is inversely related to water content of body.
c **False.** Triglycerides (neutral fats) are esters of glycerol and fatty acids.
d **True.** Sterols are classified as fats mainly because they are fat soluble.
e **False.** The more double bonds the more unsaturated a fat becomes, i.e. the fewer hydrogens the less saturated it is. A fully saturated compound has bound all the H atoms it can.

27 a **False.** Omega-3 fatty acids are long-chain, 20–22C polyunsaturated fatty acids.
b **False.** EPA and DHA have their first double bond three carbon atoms from the terminal methyl (CH_3) end.
c **False.** Linoleic acid (an 18C, omega-6 fatty acid) is converted slowly to arachidonic acid (20C) and to EPA and DHA in the liver. Linolenic acid (18C, omega-3) can do the same.
d **False.** Linoleic acid is found in soya bean oil, linseed oil, green leafy vegetables.
e **True.** Omega-3 fatty acids inhibit the hepatic synthesis of triglycerides.

28 a **False.** Chylomicrons (100–500 nm) enter the central lacteal of the intestinal villi.
b **True.** Chylomicron composition is: triglyceride 90%; phospholipid 6%; protein 2%; cholesterol 2%.
c **False.** Hormone-sensitive lipase is intracellular. It breaks down stored fat into FFA and glycerol.
d **False.** Lipoprotein lipase of endothelium is released into blood and hydrolyses the central fatty core. Products enter adipocytes.
e **True.** The liver then exports very low density lipoprotein (VLDL) which it generates from chylomicron remnants.

29 a **True.** The regular hepatic export of VLDL helps prevent a fatty liver.
b **False.** VLDL first becomes intermediate density lipoprotein (IDL).

c   **False.** Liver removes 50% of one's LDL. Other tissues, especially the macrophages, gonads, and adrenal cortex, remove the rest.

d   **True.** HDL is made in the liver and the GIT. It accepts cholesterol from all tissues via lecithin-cholesterol acyltransferase (LCAT).

e   **True.** HDL accepts cholesterol and carries it back to the liver.

30   a   **True.** About 93% of body cholesterol is in cell membranes.

b   **False.** All body cells can form cholesterol *de novo*.

c   **True.** LDL contains the highest concentration of cholesterol.

d   **False.** Cholesterol inhibits HMG-CoA. Thus, as cholesterol rises, it inhibits its own synthesis.

e   **False.** Cholesterol inhibits the production of LDL receptors. Therefore cells that use a lot of cholesterol have more LDL receptors than others.

31   a   **False.** Elevations in blood cholesterol parallel a reduction in LDL receptors. Note that this defect also occurs in familial hypercholesterolaemia.

b   **True.** Fibric acid derivatives (clofibrate, gemfibrizol) act at least partly by stimulating lipoprotein lipase.

c   **True.** Thus lovastatin inhibits the initial rate-limiting step in cholesterol synthesis.

d   **True.** LDL and also the apo B /apo E remnants of VLDL are atherogenic.

e   **True.** HDL (especially HDL2) may defer atherosclerosis by extracting cholesterol from vessels to the liver for excretion and/or by competition with LDL.

32   a   **False.** Up to recently linolenic acid was the only omega-3 fatty acid considered to be essential. The very slow biosynthesis of EPA, DHA, and arachidonic acid from linolenic and linoleic acid makes all these acids 'virtually' essential.

b   **True.** Enzymes for β-oxidation are found only in the mitochondria; enzymes for fatty acid synthesis are only in cytoplasm.

c   **False.** A single central fat droplet is typical except in brown fat.

d   **True.** Much of the glucose absorbed is converted in the liver to FFA and glycerol and exported with lipoprotein. Other sources of depot fat include chylomicrons, blood glucose, and ketogenic amino acids.

e   **False.** The number of adipocytes appears to depend in part on nutrition in early childhood.

33   a   **False.** The reserves in kcals are: fat, 140 000; protein, 24 000; and CHO, 1000.

b   **False.** Note: margarine stays soft in fridge; butter becomes hard.

c   **True.** Fat disposition in the body probably helps to prevent freezing of the most vulnerable areas.

d   **True.** As depots become saturated with anaesthetic, resistance to the anaesthetic diminishes (Gill's law of diminishing resistance).

e **True.** Metabolism of 1 g of fat yields almost twice the water that 1 g of CHO yields. Fat contains little O_2. The camel's hump is in fact a fat reserve.

34 a **False.** Alcohol is absorbed in mainly in the stomach and duodenum.

b **True.** Some of the excess H^+ that alcohol provides is accepted by pyruvate.

c **False.** However, some women appear to have more alcohol dehydrogenase than men in their stomachs but this is not the general rule.

d **False.** Alcohol is a local vasodilator but constricts via activating sympathetic nerves. It may be because it releases hypothalamic corticotrophin releasing hormone which activates the sympathetic at a central site.

e **True.** Moderate drinking may also promote coronary artery dilation. Note: there is a J-shaped relation between alcohol intake and blood pressure.

35 a **True.** Alcohol dehydrogenase facilitates alcohol acetaldehyde acetyl CoA. At both stages NAD^+ is reduced to NADH. This robs the liver of NAD^+ which is essential for gluconeogenesis.

b **False.** Approximately 8–10 g of alcohol is metabolized each hour irrespective of the blood level of alcohol.

c **True.** Alcohol appears to stimulate microsomal oxygenating enzymes so that some habitual drinkers metabolize alcohol faster than expected. This may explain the tolerance displayed by some alcoholics for other drugs metabolized in this way, e.g. barbiturates, anaesthetics, and psycotropics.

d **False.** Alcohol increases postganglionic sympathetic discharge to skeletal muscle and causes vasoconstriction here (i.e. the α-receptor effect).

e **True.** The likely mechanism is alcohol → hypothalamic CRH → nearby sympathetic centres. Vagus is also inhibited. This explains the increased cortisol release after taking alcohol.

Obesity and starvation (Qs 36–37)

36 *Concerning obesity:*

a The triceps skinfold thickness gives a practical measure of obesity.

b Obesity is said to be present when a person weighs 10% more than his or her ideal weight.

c Obesity relates more to birth weight than to weight at 5 years old.

d If two weight-matched obese people eat the same 'reducing diet' and have the same energy expenditure they will lose weight at about the same rate.

e Impaired thermogenesis may result in some people laying down fat more readily than others.

37 *Starvation*

a In a 24-hour fast more grams of endogenous fat are consumed than of protein and CHO combined.

b The basal metabolic rate is maintained despite fasting for 2 weeks.

c The brain consumption of glucose remains fairly constant even in a prolonged fast (10–30 days).

d In a long fast the increased gluconeogenesis is from all possible sources.

e In the terminal stage of long fast urinary nitrogen rises to 4 g/day.

Obesity and starvation (Answers)

36 a **True.** Triceps skinfold thickness in the 30–50 age group should be < 30 mm in women and < 25 mm in men.

b **False.** A value of 20% overweight is widely accepted as obesity. Morbid obesity is present if the body-mass index (weight/height², where weight is measured in kilogram and height in metres) exceeds 40.

c **False.** Obesity appears to relate more to weight at 5 years than at birth (see Ch7, Q52e).

d **False.** The energy density (percentage of each gram of tissue that is fat) varies considerably between people.

e **True.** For example, obesity may occur if one has defective Na^+ pumps (fat laid down to maintain body heat) or if there is a defective mobilization of adipose tissue (genetic).

37 a **False.** The average adult uses about 180 g CHO, 160 g fat, and 75 g of protein in the first day of starvation.

b **False.** The BMR falls due to a fall in total body metabolizing mass.

c **False.** Brain consumption of glucose falls from about 120 g (overnight fast) to 20 g (long fast). Ketone consumption rises to compensate.

d **False.** Increased gluconeogenesis comes mainly from glycerol. Glucogenic amino acids are conserved after 24 h.

e **False.** Urinary nitrogen falls from an initial 12–15 g/day to 3–4 g/day. In the terminal stage of fasting it rises sharply once more.

Trace nutrients (Qs 38–39)

38 *Trace dietary elements:*

a Zinc is carried in the blood combined with transferrin.

b Diets rich in whole grain and beans favour Zn absorption.

c Dietary iodine is converted to iodide in the GI tract.

d Copper has a critical role in iron metabolism.

e Fluoride in water at 2–10 p.p.m. can cause staining of tooth enamel.

39 *Vitamins:*

a Breast milk vitamin A is an important protection against nutritional blindness.

b The liver normally contains about 1 year's store of vitamin A.

c Nicotinic acid is the active form of niacin.

d Vitamin C, vitamin E, and β-carotene are all antioxidants.

e Mega doses of vitamin C are beneficial in sickle-cell anaemia.

Trace nutrients (Answers)

38 a **True.** Thus anything that binds transferrin hinders Zn absorption and carriage.

b **False.** Whole grains and beans are low in Zn and their fibre hinders Zn absorption. Meat is the richest source of Zn.

c **True.** Iodide ion occurs in food. Any traces of dietary iodine are converted to iodide. Iodine gas is poisonous.

d **True.** Copper converts ferric to ferrous iron. Copper is also needed in making collagen; thus it aids wound healing.

e **True.** Fluoride replaces the OH of hydroxyapatite. 1 p.p.m. hardens but does not stain enamel.

39 a **True.** Breast milk vitamin A protects against xerophthalmic blindness up to the third year of life. One million children go blind from xerophthalmia each year.

b **True.** Diet carotinoids are converted in the liver to vitamin A. Special transport proteins pick up vitamin A here and carry it to target cells.

c **False.** The body converts nicotinic acid to active nicotinamide. The two coenzyme forms, NAD and NADP, are essential in intermediary metabolism.

d **True.** Antioxidant vitamins may deter oxidant damage by LDL (the most atherogenic form of cholesterol).

e **False.** Mega doses of vitamin C may be harmful. Reducing agents distort HbS, may cause overt clumping of RBCs in this group.

Effects of trauma and sepsis (Qs 40–43)

40 *The ebb phase response to trauma:*

a The ebb phase is absent in uncomplicated recovery from elective surgery.

b An established ebb phase lasts up to 5 days.

c A 30-fold rise in cortisol is common during abdominal surgery.

d Most general anaesthetics provoke an ebb phase response.

e There is a rise in blood pressure, cardiac index, and O_2 consumption in the ebb phase.

41 *The flow phase:*

a   This occurs in moderate or severe trauma /infection.

b   This phase usually lasts a week or more.

c   The magnitude of the flow phase depends on the patient's age.

d   Preoperative dietary supplementation improves the subsequent flow phase.

e   The best outcomes occur when there is a good flow phase.

42 *In the flow phase:*

a   hormone-sensitive lipase is inhibited.

b   ketones increase in the plasma.

c   one should omit antidiabetic drugs or at least reduce their dosage in patients with NIDDM.

d   there is increased release of all hypothalamic releasing factors.

e   leucine donates its C skeleton for gluconeogenesis.

43 *Metabolic responses in the flow phase:*

a   The gluconeogenesis of sepsis and trauma is subject to the same control by glucose as in normal physiology.

b   Obligate loss of muscle protein is reversed by IV amino acids.

c   Large amounts of glutamine from the GIT travel to the liver and form urea and glucose.

d   Skeletal muscle is the major source of protein breakdown.

e   The BMR increases about fivefold in extreme cases.

Effects of trauma and sepsis (Answers)

40   a   **False.** Some damaged tissue, hypovolaemia, pain, and apprehension always occur. Even a skin incision causes an ebb phase.

b   **False.** The ebb phase lasts 24–48 h.

c   **True.** The ebb phase begins during the first 60 min and rises steadily. A failure of an early cortisol response may be fatal.

d   **False.** Only ether, cyclopropane, and ketamine provoke an ebb phase. The stress of induction (face mask, intubation) and other manipulations (setting up IV and assorted monitors) are more significant.

e   **True.** Part of the sympathoadrenal response to trauma/surgery.

41   a   **True.** The flow phase is a common endogenously mediated response to stimuli and toxins emanating from damaged/dead tissue and micro-organisms.

b   **True.** The duration of the flow phase is chiefly determined by the severity of the trauma.

c   **True.** The best flow phase occurs in young well-fed subjects. It is worst in the elderly and nutritionally deprived.

d **False.** Nor have special exercise training programmes improved the flow phase.

e **True.** A good flow phase is associated with rapid healing and recovery.

42 a **True.** Increased insulin inhibits hormone-sensitive lipase.

b **False.** Increased insulin in trauma and sepsis is antiketogenic.

c **False.** Despite hyperinsulinaemia there is hyperglycaemia due to an increase in counter-regulatory hormones (amines, glucagon, cortisol). Note there is also an unexplained insulin resistance in skeletal muscle.

d **False.** GnRH is inhibited in the flow phase.

e **False.** Glutamine and alanine are particularly effective gluconeogenetic agents.

43 a **False.** In sepsis and trauma glucose does not inhibit gluconeogenesis.

b **False.** IV amino acids slow but do not reverse proteolysis in the hypercatabolism of trauma. Note the fall in IGF-I despite a rise in GH in catabolic states. This suggests retention of direct anti-insulin and lipolytic actions of GH and abolition of its indirect IGF-I-mediated anabolic effects.

c **True.** Intestinal glutamine is dramatically depleted in the flow phase.

d **True.** In the full-blown hypercatabolism of trauma, glutamine levels fall by up to 50%, proline levels by about 33%. Muscle protein loss (especially collagen) varies from 100 g/day after a mastectomy to 400 g/day after multiple fractures. Note that skeletal muscle contains 45% of body protein.

e **False.** BMR increases about 1.75 times.

11

The nervous system

Autonomic nervous system (Q. 1)

1 Regarding the autonomic nervous system (ANS):

a Nuclei of origin of the sympathetic system are in the dorsal and sacral regions of the spinal cord.

b Virtually all sympathetic fibres that enter the paravertebral ganglia (including the superior cervical /stellate ganglion) synapse there.

c One presynaptic sympathetic fibre synapses with one postsynaptic one.

d It is purely a motor system.

e The postsynaptic transmitter of the sympathetic fibres is released in discrete quanta at the nerve ending.

Autonomic nervous system (Answer)

1 a **False.** The sympathetic system originates in the intermedio-lateral column from T1 to L2.

b **False.** Many preganglionic sympathetic fibres pass through ganglia unchanged to synapse in distant ganglia, e.g. the coeliac, splanchnic, and adrenal medullae.

c **False.** A preganglionic sympathetic neuron normally synapses with many postsynaptic ones. One-for-one is often found in the parasympathetic system.

d **True.** 'Autonomic afferents' and 'visceral afferents' travel with and have their cell bodies along with 'somatic afferents' in the dorsal root ganglion of spinal, or in the sensory nuclei of cranial nerves.

e **False.** NA is released along the length of the terminal axon from axonal swellings or 'varicosities'. Often no definite synaptic contacts are found.

The neuron and associated structures (Qs 2–8)

2 About the neuron:

a The soma or body lacks mitochondria.

b The soma is quite resistant to oxygen lack.

c All nerve terminals can synthesize their own neurotransmitters (NTs).

d Unipolar cells have a single process that divides into an axon and a dendrite.

e Plasticity in the adult CNS refers to the physical realignment of fibres after injury or stress.

3 In the neuron:

a Mitochondria are distributed throughout both soma and axon.

b Information from dendrites is transmitted to axon hillock by cumulative electrical changes in the plasma membrane of the cell.

c Protein synthesis occurs throughout the soma and axon.
d There is no ability to replicate.
e In the brain nutrients come directly from cerebral capillaries.

4 Within the CNS:

a Neuro-neuronal synapses are either axo-dendritic or axo-somatic.
b Axo-dendritic synapses are generally excitatory.
c Neuroglia is derived from mesoderm.
d Neuroglia contributes only a small proportion to the volume of the adult brain.
e Ependymal cells form a tight CSF–brain barrier similar to the vascular blood–brain barrier.

5 Glial cells:

a The greater part of the ependymal lining allows the free passage of water and water-soluble products in either direction.
b The pia–glial membrane is only permeable to fat-soluble solutes.
c Microcytes are the smallest and most numerous of glial cells.
d Microcytes are actively phagocytic.
e Oligodendrocytes are the functional CNS counterparts of Schwann cells.

6 Astrocytes:

a are important in supplying omega-3 fatty acids to nearby neurons in starvation.
b are important in supplying the glucose needs of nearby neurons.
c are believed to secrete K^+ into the ECF of the brain.
d provide directional cues for growing nerve fibres.
e are necessary for the uptake and recycling of neurotransmitters (NTs).

7 Concerning neural conduction:

a The nodes of Ranvier are present in all motor nerves.
b Nodes of Ranvier contain a very large concentration of Na^+ channels.
c The speed of neural conduction is faster *in vivo* than *in vitro*.
d In demyelinating conditions conduction rates are often increased.
e Compound action potentials increase as the stimulus is increased.

8 In blocking nerve conduction:

a Motor nerves are more likely to suffer than sensory ones in most nerve entrapment syndromes.
b General anaesthetics (GAs) are without effect on peripheral nerves.
c Cold inhibits conduction at different rates in different nerve fibres.
d Local anaesthetics (LAs) are usually injected in fat-soluble form for easy passage into nerves.
e Extracts of the Japanese puffer fish abolish the membrane potential.

The neuron and associated structures (Answers)

2 a **False.** The soma or body of the neuron is a very active cell.

   b **False.** The soma is extremely vulnerable to hypoxia. Unconsciousness occurs after only 8–12 s oxygen deprivation.

   c **False.** Peptide NTs are made in the cell body and transported to the nerve terminals.

   d **True.** Unipolar cells are typically found in the dorsal root ganglion of spinal nerves. The long dendrite (sensory process) here is similar to an axon.

   e **False.** Plasticity is a functional adaptation of physical linkages. Physical remodelling only occurs in the peripheral nervous system and in the immature CNS.

3 a **True.** Respiratory activity takes place both in soma and axon.

   b **False.** A mesh of very fine neurofibrils connects the dendrites to the axon hillock across the cell body.

   c **False.** Protein synthesis is confined to the cell body proper. Other metabolic processes take place throughout the entire neuron.

   d **True.** Neurons are incapable of division. The nucleus lacks a centrosome. Hence brain tumours are composed of glial tissue only.

   e **False.** Nutrition of neurons is via ECF and also probably via local neuroglia (astrocytes).

4 a **False.** Dendro-dendritic synapses are also found.

   b **True.** Axo-dendritic synapses are broad-contact (1–2 μm), large-cleft (300 A), type I synapses. They contrast with narrow-contact (1 μm), small-cleft (200 A), generally inhibitory, axo-somatic, type II contacts.

   c **False.** Neuroglia is derived from ectoderm.

   d **False.** Neuroglia contributes between 20% and 45% of the brain volume.

   e **False.** Ependymal cells are a single, copiously fenestrated layer of ciliated cells separating the CSF from the brain ECF.

5 a **True.** The ependymal lining is permeable except over the choroid plexus, where the cells are linked by tight junctions.

   b **False.** The pia–glial membrane is highly hydrophilic. Note that the pia is derived from the mesoderm.

   c **False.** Microcytes are the least numerous of the glial cells. They function mainly in inflammation and repair.

   d **True.** Phagocytosis is the main specific function of the microcyte.

   e **True.** Oligodendrocytes are needed for the formation and repair of myelin sheaths. They are dysfunctional in demyelinating diseases such as multiple sclerosis.

6 a **False.** Neurons do not use fatty acids for energy, even in starvation.

   b **True.** Astrocytes have a special role in supplying nearby neurons with glucose.

c **True.** Astrocytes help maintain K^+ levels in brain ECF primarily by mopping up K^+ ions. However, they can also make K^+ available to nearby neurons.

d **True.** Astrocytes are critical for the organizational pattern in the developing nervous system. They are also critical for maintaining the blood–brain barrier.

e **True.** Astrocytes are necessary for the uptake and recycling of NTs, especially of γ-aminobutyric acid (GABA) and glutamate.

7 a **False.** Nodes of Ranvier are found only in myelinated nerves.

b **True.** Nodes of Ranvier are rich in Na^+ channels and contain about $10\,000/m^2$.

c **False.** *In vivo* neural conduction rates are generally much slower than *in vitro* rates.

d **False.** Thinner myelin allows Na^+ channels to spread out from the nodes, thus slowing conduction rates. Thus saltatory conduction is lost.

e **True.** Because of the different thresholds of the different nerve fibres when whole nerves are stimulated to give a compound action potential. There can be a gradual increase in size of the potential up to a maximum when all fibres have been recruited.

8 a **False.** Sensory nerves are more vulnerable to musculofascial entrapment, causing parasthesiae, pain, and numbness.

b **False.** GAs block conduction by interfering with cation flux, by inducing chronic depolarization, by altering synaptic transmission, by expanding lipids in bilayer (critical volume hypothesis), or by any combination of these.

c **True.** Cold (and ischaemia) block conduction by reducing local metabolism. Individual fibres vary in their resistance.

d **False.** LAs are normally ionized and water soluble. They are converted to hydrophobic non-ionized forms in the body.

e **False.** Tetrodotoxin blocks Na^+ channels by blocking initiation or propagation of the action potential (AP).

Cerebrospinal fluid (Qs 9–11)

9 *In the secretion and absorption of cerebrospinal fluid (CSF):*

a CSF is actively secreted by the endothelial cells of the choroid plexus.

b There are tight junctions between the ependymal cells that line the choroid plexuses.

c The vascular endothelial cells in the chorioid plexus are loosely bound.

d Absorption in venous sinuses is strictly unidirectional in normal people.

e Some absorption of CSF is carried out by CNS neurons.

10 *Volume and composition of CSF:*

a   About 1.0 ml is formed per minute.
b   About half of the CSF is in the ventricles at any one time.
c   CSF has a similar composition to plasma except for protein.
d   CSF protein levels are about half that in plasma.
e   CSF glucose falls dramatically in tuberculous meningitis.

11 *The CSF:*

a   helps maintain a constant intracranial volume.
b   removes waste products of cerebral metabolism.
c   has a critical controlling role on important brain functions.
d   has a nutritive role for the brain.
e   forms an hydraulic cushion, almost halving the virtual weight of the brain.

Cerebrospinal fluid (Answers)

9   a   **False.** CSF-secreting ependymal cells cover the choroid plexuses in the lateral ventricles and third ventricle (see 9c).
    b   **True.** This contrasts with ependymal cells elsewhere.
    c   **True.** Loosely bound vascular cells allow an ultrafiltrate of plasma to form between themselves and ependymal cells.
    d   **True.** A CSF–venous pressure gradient of 50 mmHg is needed for the proper absorption of CSF.
    e   **False.** A small amount of CSF is reabsorbed by lymphatics that run in the sheaths of the spinal nerves.

10 a   **False.** About 0.5 ml of CSF is formed per minute. Adult volume is approximately 120–140 ml of which 70% comes from the choroid plexuses and 30% from endothelial cells lining brain capillaries.
    b   **False.** The distribution of CSF is 22 ml in the ventricles; 45 ml in the cranial subarachnoid space; and the rest in the spinal subarachnoid space and central canal.
    c   **False.** CSF has less glucose (2.5–4.5; i.e. 2/3 of blood levels), bicarbonate (22.0), calcium (1.1), potassium (3.5), and more chloride (120–130) than plasma . All values are expressed in mmol/l.
    d   **False.** CSF is virtually protein free (only 0.2–0.4 g/l).
    e   **False.** TB meningitis is noted for a dramatic fall in CSF chloride.

11 a   **True.** In conditions producing cerebral atrophy there is an increase in CSF production, and with cerebral swellng CSF production diminishes.
    b   **True.** CSF is a natural draining channel for excess ECF and waste produce of cerebral metabolism. CSF is continuous with the ECF of the brain.
    c   **True.** CSF regulates brain function in many ways, e.g. CSF pH and PCO_2 on respiration; CSF NaCl on AVP; and CSF carriage of neuromodulators

e.g. opioids. Complete anaesthesia occurs when the when pH of CSF falls to 6.7

d **True.** The CSF probably plays a minor role in CNS nutrition, e.g. in supplying glucose to inaccessible areas.

e **False.** The CSF reduces a 1400 kg brain to a virtual weight of only 40 g, thus protecting vital neurons from being crushed.

Sensory nerve endings (Q. 12)

12 In the functioning of sensory nerve endings:

a There is an overlap between the specific receptors for different modalities (pain, temperature, touch, vibration, and pressure).

b Free nerve endings are found innervating individual cells.

c Free nerve endings are only found in the skin and surface epithelium.

d Meissner's corpuscles are embedded in onion-like layers of connective tissue found in dermal papillae.

e Sub-threshold stimuli exert a change in the resting membrane potential of sensory nerve endings.

Sensory nerve endings (Answer)

12 a **True.** The specificity of receptors for individual modalities is not now regarded as absolute.

b **False.** Free never endings are found lying between individual cells.

c **False.** In deeper tissues layers, free nerve endings innervate muscle spindles, vascular smooth muscle, and chemosensitive cells that subserve stretch and the local chemical environment.

d **False.** Meissner's corpuscles are encapsulated spiral dendritic found in dermal papillae sensitive to light touch. Pacinian corpuscles are onion-like organs found in connective tissue and respond to changes in pressure and vibration.

e **True.** Subliminal stimuli will not set up a full-blown AP, but may set up small generator potentials. They can change the membrane potential but do not cause changes that reach the threshold for a stimulus to be conveyed.

Neural transmission and neural transmitters (Qs 13–35)

13 In chemical synaptic transmission:

a All synaptic clefts have much the same width.

b Presynaptic vesicles only discharge into the cleft after an AP has arrived at the presynaptic terminal.

c Arrival of an AP at the presynaptic membrane allows Ca^{2+} to enter the cleft.
d NTs such as ACh, NA, and amino acids are made in axon terminals.
e Excitation is usually associated with the net passage of cation from cleft into the postsynaptic cell.

14 Concerning electrical transmission:

a As far as is known, all electrical synapses are excitatory.
b Electrical transmission is faster than chemical transmission.
c In humans electrical transmission only occurs in the brain.
d Electrical transmission is more useful than chemical transmission in cold-blooded animals.
e Ephaptic transmission (ephapsis, cross-talk) is due to ionic movement through gap junctions.

15 Changes in the postsynaptic membrane:

a Once there is a net transfer of cations from the cleft into the subsynaptic cytoplasm then a propagated AP is inevitable.
b An excitatory postsynaptic potential (EPSP) only lasts 1–2 ms.
c In polysynaptic input (e.g. motorneuron in spinal cord) even an excess of one EPSP over inhibitory potentials (IPSPs) can produce a spike potential (AP) at the axon hillock.
d In the skeletal muscle end plate IPSPs exceed EPSPs when the muscle is at rest.
e Sensitivity to EPSPs is greater the greater the polarization of the membrane.

16 Following changes in membrane potential caused by transmitters:

a In a neuron incoming impulses depolarize the axon hillock without depolarizing the intervening soma.
b A given excitatory postsynaptic potential (EPSP) causes a greater change at smaller than at larger motorneurons.
c Different neurotransmitters (NTs) cause EPSPs and IPSPs (inhibitory postsynaptic potentials).
d γ-Aminobutyric acid (GABA) causes presynaptic inhibition (blocks Ca^{2+} entry into presynaptic terminal (see Q 27c).
e Presynaptic inhibition is very important in damping down visceral and somatic afferent activity in the spinal cord.

17 Postsynaptic inhibition:

a is due to hyperpolarizaton of the sub and/or postsynaptic membrane.
b is facilitated by the amino acid alanine.
c is facilitated by strychnine.
d occurs at recurrent motorneuron axon/Renshaw cell synapse.
e occurs at Renshaw cell/Renshaw cell synapses.

18 *In cholinergic transmission:*

   a Choline for acetylcholine can be made *de novo* in nerves.

   b About 50% of released ACh is taken back intact into the nerve terminal.

   c α-Bungarotoxin (which combines with ACh receptors) simulates the effects of ACh at the motor end plate.

   d The effects of ACh are mimicked by applying ACh directly to skeletal muscle sarcolemma.

   e Individual cholinergic synapses contain both muscarinic and nicotinic receptors.

19 *Muscarinic (M) and nicotinic (N) receptors:*

   a In general M_1-receptors are excitatory and M_2-receptors are inhibitory.

   b Postsynaptic M- and N-receptors mediate fast EPSPs.

   c Muscarinic receptors outnumber nicotinic ones by over 100 to 1 in the CNS.

   d M_3-receptors are stimulatory on both pre- and postjunctional tissue.

   e Most excitatory CNS synapses are cholinergic in nature.

20 *In myasthenia gravis:*

   a The number of ACh molecules released at the neuromuscular junction in skeletal muscle is dramatically reduced.

   b The functional neuromuscular abnormality is due to antibodies against the ACh receptor.

   c There is a normal histological neuromuscular junction.

   d The condition does not occur in the newborn.

   e Steroids can benefit up to 90% of cases.

21 *In adrenergic transmission:*

   a In the CNS only about 1% of neurons are adrenergic.

   b Amine discharge, like that of ACh, requires the entry of Ca^{2+} into the presynaptic terminal.

   c NA is stored in preformed vesicles in sympathetic terminals.

   d Up to half of the secreted NA is recycled by the nerve terminal.

   e α1-Adrenoreceptors are about equally distributed on pre- and postsynaptic membranes.

22 *Non-β aminergic transmission:*

   a α2-receptors are found on the presynaptic terminals of both adrenergic and cholinergic synapses.

   b α2-receptors generally work by increasing the activity of adenylate cyclase.

   c Cotransmission involves simultaneous discharge of sympathetic and parasympathetic neurons in same organ.

   d Non-adrenergic non-cholinergic (NANC) transmission is a feature of many parasympathetic plexuses.

   e Acting as an NT the effect of DOPA appears to be confined to the CNS.

23 In the pharmacological manipulation of the autonomic nervous system:

a   No known drug inhibits the formation of ACh in cholinergic terminals.
b   All of the enzymes involved in the biosynthesis of catecholamines can be inhibited by known drugs.
c   Botulinum toxin blocks ACh receptors.
d   Cocaine blocks amine reuptake by sympathetic nerves.
e   Numerous drugs inhibit the breakdown of NTs in the sympathetic system.

24 Regarding adrenergic drugs:

a   Sympathomimetic amines (ephedrine, tyramine, and amphetamine) bind to adrenergic receptors.
b   Prazosin blocks α 1-receptors.
c   Clonidine and dexmedetomidine inhibit α 2-receptors.
d   Isoproterenol (isoprenaline) stimulates both β 1- and β 2-receptors.
e   Cardioselective β blockers block β 2-receptors.

25 In neuromuscular (NM) blockade:

a   Non-depolarizing blockers are extremely lipid soluble.
b   Non-depolarizing agents act by mimicking the effect of ACh at the end plate.
c   Depolarizing agents are not as easily destroyed as ACh at the motor end plate.
d   Atropine causes a non-depolarizing blockade at the skeletal NM junction.
e   Local anaesthetics (LAs) with a pK close to body pH have fastest onset of action at the NM junction.

26 Transmission within the CNS:

a   The locus coeruleus (LC) is an important aminergic area in the CNS.
b   Cholinergic transmission is important in the ascending reticular activating system (ARAS).
c   Neutral amino acids are generally excitatory in the CNS.
d   Dicarboxylic amino acids, e.g. homocysteic acid, are neither excitatory nor inhibitory.
e   Adenosine is a natural anticonvulsant neurotransmitter.

27 GABAergic transmission in the CNS:

a   GABA is formed in the liver and transported to the CNS where it is stored and later released by GABAergic neurons.
b   GABA is stored in vesicles within nerve terminals.
c   GABA exerts at least part of its effect by closing Cl$^-$ channels.
d   GABA accounts for about 1% of CNS transmission.
e   The GABA receptor is co-located with the benzodiazepine receptor.

28 Clinical relevance of GABA:

a   Benzodiazepines enhance the effect of GABA on chloride channels.

b   GABA stimulates dopaminergic cells in Parkinson's disease.

c   Hepatic encephalopathy is associated with decreased GABA activity.

d   Tetanus toxin inhibits GABA release.

e   A decrease in the activity of glutamic acid decarboxylase (GAD) results in a flaccid paralysis.

29 About dicarboxylic amino acid transmitters:

a   In the CNS the highest concentration of L-glutamic acid and aspartic acid is found in the brain stem.

b   Dicarboxylic acids excite by increasing Na^+ entry through the postsynaptic membrane.

c   Glutamic acid is a precursor of GABA.

d   Brain glutamic acid originates in the liver.

e   There is far more free glutamate in brain ECF than in brain cells.

30 Excitotoxicity:

a   Dietary glutamate can cross the blood–brain barrier.

b   Glutamate has been specifically implicated in memory control.

c   NMDA receptors (N-methyl-D-aspartate receptors) allow ingress of Ca^{2+} into brain cells.

d   Astrocytes dampen excitation by controlling ECF glutamate.

e NO^+ (nitrosonium ion) stimulates NMDA receptors.

31 Dopaminergic transmission:

a   The ability to form dopamine (DA) is best developed in the CNS.

b   Some neurons of the nigrostriatal tract form and release NA.

c   In general the actions of DA are inhibitory in the CNS.

d   Dopamine is poorly recycled by dopaminergic terminals.

e   Both D_1- and D_2-receptors act via adenylate cyclase.

32 Serotoninergic transmission (5-HT):

a   Most of the body 5-HT is in the CNS.

b   5-HT readily passes the blood–brain barrier.

c $5\text{-}HT_1$ receptors relax smooth muscle (e.g. gut, blood vessels, and the uterus).

d $5\text{-}HT_2$ receptors cause vasoconstriction and bronchospasm.

e $5\text{-}HT_3$ receptors are found mainly on nerves and are both inhibitory and excitatory in nature.

33 About serotonin (5-HT) in the CNS:

a   The $5\text{-}HT_1$ receptor is the chief 5-HT receptor in the brain.

b   CNS serotonin is mostly involved in basal ganglia functions and in co-ordination activity.

c   Monoamine oxidase inhibitors (MAOIs) increase CNS 5-HT, NA, DA.
d   Inhibition of tryptophan hydroxylase causes insomnia.
e   Reserpine disrupts the storage of 5-HT in synaptic vesicles.

34 *Manipulating serotonin in the CNS:*

a   The anorectic effect of fenfluramine is mainly due to release of 5-HT from its vesicle stores.
b   Depletion of 5-HT stores can cause depression.
c   Inhibition of the postjunctional uptake of 5-HT (and of NA) is helpful in many cases of depression.
d   Tricyclic antidepressants block muscarinic receptors.
e   The cytochrome P450 enzyme is a major target of newer antidepressant drugs, e.g. fluoxetine, sertraline, and paroxetine.

35 *Regarding the release of neurotransmitters in the CNS:*

a   Methylenedioxymethylamphetamine ('Ecstasy') interferes with the normal glutamate: GABA ratio in the brain.
b   Peptide NTs are made and packaged in the cell body.
c   Substance P is released along with glycine by Renshaw cells.
d   Many peptide NTs act as cotransmitters in the CNS.
e   All of the following are peptide neurosecretions or peptide NTs in the CNS: VIP, tyramine, somatostatin, vasopressin, oxytocin, and tryptamine.

Neural transmission and neural transmitters (Answers)

13   a   **False.** Synaptic clefts vary from about 20 nm to as little as 2 nm.
b   **False.** Random discharge occurs in the resting state. It is indexed by a low-level electrical activity recorded at the subsynaptic membrane.
c   **False.** AP opens up voltage-gated channels and allows Ca^{2+} to enter the presynaptic terminal to facilitate vesicle discharge.
d   **True.** Peptides, enzymes, vesicles, and organelles are made in the cell body.
e   **True.** Conversely, inhibitory conductance changes facilitate Cl^- entry and/or increase K^+ exit from the postsynaptic cell.

14   a   **True.** This curtails their versatility considerably *vis-à-vis* chemical transmission.
b   **True.** This is because no chemical-change time delay is involved.
c   **False.** Electrical transmission occurs in smooth and cardiac muscle cells where cell boundaries have virtually the same electrical properties as the cytoplasm.
d   **True.** This is because electrical transmission is independent of temperature changes.

e **True.** Ephapsis is probably confined to the CNS. A rise in ICF Ca^{2+} closes these tiny channels. Unwanted ephapsis may cause bizarre symptoms in nerve injury and in multiple sclerosis.

15 a **False.** A small transfer of ions causes a non-propagated excitatory or inhibitory postsynaptic potential (EPSP or IPSP).

b **True.** EPSPs last 1–2 ms and destabilize the membrane potential by –5 mv to –15 mv.

c **False.** It needs a ratio of 100:1 in order to produce an AP. Remember, a neuron may have 6000 synapses impinging on its body.

d **False.** There are no IPSPs at the motor end plate.

e **True.** Thus at three membrane potentials of –95 mV, –70 mV, and –45 mV the corresponding EPSPs are –15 mV, –10 mV, and –5 mV.

16 a **True.** The axon hillock is rich in Na^+ channels and thus easily depolarized; the rest of the soma has few channels.

b **True.** This means that small motorneurons are the easiest excited.

c **False.** EPSP or IPSP is ultimately a property of the subsynaptic membrane, irrespective of which type of NT is released.

d **True.** GABA also exerts postsynaptic inhibition (see Answer 27c). Picrotoxin and bicuccilin inhibit GABA.

e **True.** Presynaptic inhibition smooths out sensory information and allows a more measured motor response.

17 a **True.** Postsynaptic inhibition is accompanied by electrophysiologically recordable IPSPs.

b **False.** Glycine, not alanine, is a powerful postsynaptic inhibitor.

c **False.** Even a few milligrams of strychinine block postsynaptic inhibition by antagonizing glycine.

d **False.** ACh is released by recurrent motorneuron axons. This stimulates the Renshaw cell to release glycine on its partner motorneurons.

e **True.** Renshaw cells inhibit other Renshaw cells, other motorneurons, and a variety of inhibitory internuncials. Alcohol stimulates Renshaw cells.

18 a **False.** The neurilemma, especially of nerve terminals, picks up choline from ECF.

b **True.** The remaining 50% of released ACh is destroyed by plasma cholinesterase.

c **False.** Bungarotoxin, a snake venom toxin, combines irreversibly with ACh receptors and blocks the normal effects of ACh.

d **False.** ACh receptors are only located on the outer surface of the end plate in the junctional folds. Application elsewhere or intracellular injection is without effect.

e **True.** The ultimate response depends on the numerical dominance of one type of receptor over the other.

19 a **True.** Furthermore, M_1-receptors are only on postsynaptic membranes, M_2-receptors are found on both.

   b **False.** N-receptors mediate fast EPSPs; M-receptors mediate slow EPSPs.

   c **True.** In general, M-receptors cause EPSPs by decreasing K^+ conductance and cause IPSPs by increasing K^+ conductance.

   d **False.** M_3 is excitatory on postjunctional and autoinhibitory on prejunctional tissue.

   e **False.** Only about 10% of CNS synapses are cholinergic.

20 a **False.** In myasthenia there is a normal synchronous discharge of 100–220 vesicles. Each vesicle, or quantum, contains about 100 000 molecules of ACh.

   b **True.** Antibodies against the ACh receptor are almost certainly the cause of myasthenia, although only 90% of patients have them in their serum.

   c **False.** In myasthenia the synaptic cleft is widened and the synaptic folds are very shallow.

   d **False.** Transient neonatal myasthenia is due to placental transfer of Ig. This occurs in about 1 in 6 myasthenic mothers.

   e **True.** Steroids help in myasthenia because they are immunosuppressants (only 30% of patients still respond after 5 years of therapy). Cholinesterase inhibitors and other immunosuppressants such as azathioprine, plasma exchange, and sometimes thymectomy are also used.

21 a **True.** Adrenergic includes NA release.

   b **True.** The entry of Ca^{2+} into the presynaptic terminal is an almost a universal prerequisite for NT release.

   c **True.** NA-filled vesicles are also stored in bulbous 'varicosities' along the course of the terminal nerve fibre.

   d **False.** There is reuptake of 80–85% of secreted NA. The same occurs with DOPA.

   e **False.** α 1-adrenoreceptors occur predominantly on the postsynaptic membrane and are mainly excitatory.

22 a **True.** In general, α 2-receptors are autoinhibitory.

   b **False.** α 2-Receptor activity generally depresses adenylate cyclase.

   c **False.** Cotransmission involves more than one NT from the same terminal, e.g. NA and VIP from sympathetic nerves at eccrine glands.

   d **True.** See NANC transmission in smooth muscle of gut and bronchi (Ch 7 Q7c; Ch 6Q21b). Also note that there is NANC activity in the pelvic ganglia of the sympathetic system.

   e **False.** Evidence also exists that DOPA is a NT in autonomic ganglia.

23 a **False.** Hemicholinium blocks the synthesis of ACh by interfering with the uptake of choline.

   b **True.** For example, carbidopa blocks DOPA dopamine; fusaric acid blocks dopamine to noradrenaline.

c **False.** Botulinum toxin attaches to cholinergic terminals, is internalized and blocks ACh exocytosis (see therapeutic uses ch 7 Q7e).

d **True.** Tricyclic antidepressants e.g. imipramine block amine reuptake.

e **True.** Clinical monoamine oxidase inhibitors (MAOIs) are very common. Pargyline and tranylcypromine also inhibit MAO; catechols inhibit COMT (catechol ortho methyltransferase).

24 a **False.** Sympathomimetics enter sympathetic fibres and displace NA, which then diffuses out to bind to adrenergic receptors.

b **True.** Yohimbine blocks α 2-receptors; phenoxybenzamine, phentolamine, and tolazoline block α 1- and α 2-receptors.

c **False.** Clonidine and dexmedetomidine stimulate α 2-receptors on the prejunctional membrane. This inhibits the further release of NT.

d **True.** Phenylephrine stimulates α 1-receptors; dobutamine stimulates β 1-receptors.

e **False.** Examples of 'cardioselective' blockers include metoprolol and atenolol, which block β 1-receptors.

25 a **False.** Non-depolarizing agents are water soluble, ionized, and contain at least two charged N atoms. Examples are tubocurarine, gallamine, alcuronium, pancuronium, and atacurium.

b **False.** Non-depolarizing agents compete with ACh for nicotinic receptors on the postsynaptic membrane.

c **True.** Depolarizing agents include true cholinergics such as suxamethonium, anticholinesterases such as neostigmine (not suitable as muscle relaxants because of widespread effects on ANS), and nicotine-like drugs such as lobeline (many side-effects).

d **False.** Atropine is a muscarinic blocker and is ineffective at ganglia or at the motor end plate (nicotinic).

e **True.** This occurs as a major portion of the LA then exists in the uncharged (diffusible) form. LAs act both at the NM junction and along nerve processes.

26 a **True.** The LC of the mid-brain is the origin of NA fibres that travel up to the neocortex and down to the brain stem and cord.

b **True.** The cells of origin of the cholinergic fibres that impinge on the reticular formation are concentrated in the basal ganglia, hippocampus, and reticular nuclei. They are mainly excitatory.

c **False.** GABA, glycine, and taurine are depressant NTs.

d **False.** Glutamic, aspartic, cysteic, and homocysteic acids are excitatory NTs.

e **True.** Much evidence supports an anticonvulsant role for adenosine. Note that adenosine has depressant effects on the heart.

27 a **False.** GABA is formed in the CNS by the decarboxylation of L-glutamic acid in nerve terminals.

b **True.** Glutamic acid decarboxylase (GAD) is found in nerve terminals. Pyridoxal phosphate (vitamin B_6) is a cofactor.

c **False.** GABA opens Cl^- channels linked to GABA A receptors. The GABA inhibition effect is slow (about 1.0 s).

d **False.** 30–40% of CNS synapses are estimated to be GABAergic. Although abundant in the substantia gelatinosa, most GABA synapses are located above cord level. GABA is the most common NT.

e **True.** This suggests that diazepine-like substances are produced naturally.

28 a **True.** A GABA-like effect has been postulated as part of the anxiolytic action of benzodiazepines in the limbic system.

b **False.** GABA inhibits dopamine via the striatonigral tract. GABA is not helpful in Parkinson's disease.

c **False.** Increased GABAergic activity may play a part in the onset of hepatic coma. Flumazenil (a benzodiazepine receptor antagonist) may lighten coma.

d **True.** Picrotoxin blocks Cl^- channel, bicuculline blocks GABA receptor, hydrazides block GABA synthesis. Hence symptoms of tetanus are unopposed α-motor fibre activity.

e **False.** An autoantibody against GAD causes the 'stiff man syndrome' (note: GAD synthesizes GABA).

29 a **False.** The highest concentration of dicarboxylic acids (DCAs) is in the cerebral and cerebellar cortices.

b **True.** Contrast DCA effect with that of GABA and glycine which open Cl^- channels.

c **True.** Glutamic acid:GABA ratio helps determine CNS excitability.

d **False.** Glutamic acid is mainly made in muscle and brain.

e **False.** CNS glutamate: ICF 10 mmol/l; ECF < 0.6 mol/l.

30 a **True.** Domoic toxin (seaweed poisoning) and glutamate analogues (Chinese restaurant syndrome) cause CNS symptoms. Note that most cases of diarrhea and vomiting after take-away foods is caused by the spore-bearing organism *Bacillus cereus* which is found in rice reheated in large pots that are not regularly cleaned.

b **True.** Glutamate is important in memory, cognition, movement, and in sensation perception.

c **True.** NMDA receptors are coupled to Ca^{2+} and Na^+ channels. Amino proprionate, kainate, quisqualate, and metabotropic receptors also respond to glutamate.

d **True.** In brain damage from any cause astrocytes lose glutamate to the ECF. Note astrocytes have a vital role in creating and in maintaining the blood–brain barrier.

e **False.** NO^+ decreases activity of NMDA receptors; in contrast, excess NO may stimulate NMDA and cause neuronal death (see ch 13 Q15d).

31   a   **False.** DOPA is formed with equal facility outside the CNS.
   b   **False.** The pathway from phenylalanine stops short at the DOPA stage.
   c   **True.** Dopamine inhibits cholinergic neurons in the striate body and in the retina. Dopamine also inhibits the release of prolactin (PRL).
   d   **False.** Dopamine is highly efficiently recycled, like NA and 5-HT.
   e   **True.** D_1-receptors stimulate adenyl cyclase; D_2-receptors inhibit adenyl cyclase.

32   a   **False.** Only about 2% of the body 5-HT is in the CNS.
   b   **False.** L-Tryptophan crosses the blood–brain barrier by an active mechanism. The CNS must make its own 5-HT.
   c   **True.** 5-HT_1 receptors are blocked by methysergide.
   d   **True.** 5-HT_2 receptors are blocked by ketanserin.
   e   **True.** 5-HT_3 are often found on presynaptic nerve terminals where they either facilitate or inhibit activity. Antagonists include MDL 72222 and GR 38032.

33   a   **False.** The 5-HT_2 postsynaptic receptor is probably responsible for most of the CNS effects of 5-HT (inhibition/excitation).
   b   **False.** CNS 5-HT appears to be mainly involved in appetite, sex, aggression, emotion, mood, impulse control, pain, sleep, and thermogenesis. It is found mainly in cells of brain stem, limbic system, and raphe nuclei.
   c   **True.** The most marked effect of the MAOIs is on 5-HT. MAOIs are useful in the treatment of depression and anxiety.
   d   **True.** Parachlorphenylalanine (PCPA) inhibits tryptophan hydroxylase and so reduces 5-HT production in the brain. This results in alertness and insomnia.
   e   **True.** The effects of reserpine are complicated. In general it depletes amine stores and may act as a false neurotransmitter. Altered sleep pattern and altered mood are features.

34   a   **True.** Fenfluramine stimulates the release of 5-HT from nerve terminals in the satiety centre. This is the basis of its anorectic effect.
   b   **True.** Fenfluramine and related anorectics, e.g. para-chloramphetamine, also block 5-HT reuptake and so deplete 5-HT stores.
   c   **False.** Inhibition of the prejunctional reuptake of 5-HT and NA appears to be of significant benefit in depression.
   d   **True.** Imipramine, desipramine, clomipramine, triptylines, and doxepin also block H_1 histaminergic and α 1-adrenergic receptors. Hence side-effects such as dry mouth, sedation, and hypotension are common.
   e   **True.** P450 has major responsibilty for the elimination of tricyclic antidepressants, selective 5-HT reuptake inhibitors and many β-blockers. While P450 is cheifly found in the liver, a similar enzyme is also found in the brain.

35   a   **False.** Ecstasy and other hallucinogens (such as lysergic acid diethylamide (LSD), mescaline, and indolamines) mimic/distort the metabolism of 5-HT and other amines (e.g. the α-adrenergic effects of Ecstasy cause urinary retention).

b   **True.** Also peptide NTs are transported to the terminals by axonal transport.

c   **False.** Substance P is released by primary afferent pain fibres in the dorsal horn. It coexists with 5-HT in the brain.

d   **True.** The role of many cotransmitted peptide NTs is often unclear. It may be a function of memory and behaviour modulation.

e   **False.** Tyramine and tryptamine are trace amine NTs in the CNS.

Opioids (Qs 36–37)

36 *Opioid peptides:*

a   Encephalins are widely distributed in the CNS.

b   Dynorphins are located mainly in the pituitary with enkephalins.

c   Endorphins are secreted by the same neurons as other opioids.

d   The μ-receptor mediates spinal analgesia.

e   The κ-receptor is stimulated by pentazocine.

37 *Opioids and opiates:*

a   Opioids and opiates are synonymous terms.

b   Each class of endogenous opioid is derived from a different independent precursor.

c   Opioid inhibition of pain is quiescent under quiet physiological conditions.

d   Autonomic reflexes (respiration, CVS) are under tonic opioid influence in normal resting states.

e   Opiate receptors are synthesized in the dorsal root ganglion and are transported both centrally and peripherally.

Opioids (Answers)

36   a   **True.** Encephalins are widely distributed in the CNS and are especially found in the basal ganglia and substantia gelatinosa. They are penta-peptides and bind to δ-receptors.

b   **False.** Dynorphins are located predominantly in the pituitary with β- endorphins.

c   **False.** Endorphins are released alongside ACTH. The same gene encodes ACTH, MSH, and endorphins.

d   **False.** The μ-receptor probably mediates central analgesia, respiratory depression, euphoria, and addiction. It is stimulated by β-endorphins and morphine.

e   **True.** Pentazocine stimulates the κ-receptor. Dynorphin is an endogenous stimulator. Both mediate spinal analgesia but not respiratory depression.

37 a **False.** Opiates include anything derived from the juice of the poppy (or are loosely related to morphine). Opioids are any directly acting compounds, the effects of which are stereoscopically antagonized by naloxone.

  b **True.** Proencephalin A is derived from encephalins, prodynorphin from dynorphin, and promelanocortin from endorphins, ACTH and β- lipoprotein.

  c **True.** There is no significant tonic secretion of opioid analgesics. The system is only activated in pathophysiology/stress.

  d **False.** Opioids become important in modulating autonomic reflexes in stress (such as hypoxia, shock, anaesthesia, pain, and neonatal state).

  e **True.** Thus opiates exert both central and peripheral effects on nerves.

Reflexes (Qs 38–41)

38 In reflex actions:

  a The interval between the stimulus and the final effect is chiefly determined by the length of the afferent and efferent nerves.

  b Spatial summation refers to simultaneous threshold stimulation of two or more nearby afferent nerves.

  c Synapses are the first part of a reflex to become fatigued.

  d Flexor reflexes are probably the most primitive of all reflexes.

  e The Babinski response is an example of a flexor reflex.

39 Regarding reflexes:

  a Tonic neck reflexes cause characteristic movements of the limbs in response to movements of the head.

  b The righting reflexes include reflexes initiated by the action of gravity on the otolith organ.

  c The red nucleus is very important in integrating postural reflexes in man.

  d Superficial abdominal reflexes involve transmission via the spinal cord.

  e The vagus (X) nerve is the afferent pathway for the gag reflex.

40 The stretch reflex:

  a It does not normally respond to very minor degrees of stretch.

  b The receptor organ is the extrafusal muscle fibre.

  c The muscle spindles involved in this reflex are most plentiful in the large antigravity muscles.

  d Nuclear chain-type fibres respond best to sharp, rapid, phasic stretch.

  e The γ-efferent fibres that supply the spindles are unmyelinated.

41 In musculo-tendinous proprioception:

  a γ- Efferent fibres comprise about 30% of an average anterior root.

  b γ- Efferent activity only takes place in stretched muscle.

  c Muscle spindles are essentially tension monitors in muscle.

d   Golgi tendon organs respond to muscle vibration.

e   Golgi tendon organs are silent when a muscle is at rest.

Reflexes (Answers)

38   a   **False.** Reflex time is chiefly determined by synaptic delay. It may be as little as 0.2–0.5 ms in a monosynaptic reflex.

b   **False.** Summation of subthreshold stimulation is involved in spatial summation within one neuron.

c   **True.** Nerve fibres are virtually indefatigable.

d   **True.** Flexor reflexes are protective in purpose, withdrawal in character, and are provoked by potentially nocioceptive stimuli.

e   **True.** The Babinski response is seen in infants. It is also seen in deep sleep, coma, general anaesthesia, and in an 'upper motor neuron lesion'.

39   a   **True.** Tonic neck reflexes are elicited by stretching proprioceptors in the neck muscles and in the joints of the cervical vertebrae.

b   **True.** Part of a sequence of reflexes is aimed at keeping the head-up position in relation to the rest of the body.

c   **False.** The red nucleus is of little significance in man.

d   **True.** All superficial abdominal reflexes have centres in the spinal cord from T6 to T12.

e   **False.** For the gag reflex: the afferent is the IXth nerve (from the lateral wall of the pharynx); and the efferent is IX to stylopharyngeus and X to pharyngeal constrictors.

40   a   **False.** The stretch reflex is very sensitive. It responds to a stretch of only 0.05 mm in 1/20 s.

b   **False.** The receptor organ is the intrafusal muscle fibre (muscle spindle).

c   **False.** Muscle spindles are most plentiful in the muscles involved in precise movements, e.g. ocular, pharyngeal, hand.

d   **False.** Nuclear bag fibres respond best to rapid stretch. Chain-type fibres respond best to continuous or tonic stretch.

e   **False.** γ-Efferent fibres are finely myelinated.

41   a   **True.** Also the cell bodies of γ-efferents lie alongside those of ordinary motorneurons in the anterior horn.

b   **False.** γ-Efferents are also active in muscle at normal length and even in shortened muscle.

c   **False.** Muscle spindles are length monitors. They lie parallel to the extrafusal muscle fibres.

d   **True.** 20–30 muscle fibres are inserted into a single Golgi tendon organ.

e   **True.** Golgi tendon organs are tension sensors. They prevent the overstretching of muscle and inhibit the motorneurons of the extrafusal compartment. They are more sensitie to active muscle sensitive/contraction than to passive muscle stretch.

Tracts in the cord (Qs 42–45)

42 *In descending tracts in the spinal cord:*

a The lateral corticospinal tract extends laterally to the surface of the spinal cord.

b The vestibulospinal tract is a major crossed tract from the opposite vestibular nuclei.

c The vestibulospinal tract predominantly inhibits extensor motorneurons.

d Reticulospinal fibres are scattered throughout the anterior white columns.

e Reticulospinal fibres predominantly excite flexor motorneurons.

43 *In the ascending tracts in the spinal cord:*

a The fasciculus gracilis and cuneatus contain fibres that mediate tactile discrimination.

b The lateral spinothalamic tract carries crude touch and pressure.

c The tract of Lissauer is made up of second-order sensory neurons.

d The spinocerebellar tracts convey impulses from Golgi tendon organs.

e All afferent fibres cross the mid line at some stage in the spinal cord.

44 *After injury to the spinal cord:*

a Spinal transection is much more serious in the thoracolumbar than in the cervical cord.

b In spinal shock urinary incontinence is due to inactivity of α 1-receptor-mediated effects on bladder.

c Flexor somatic reflexes are the first to return after spinal shock.

d Whiplash injury is almost certainly a psychological entity.

e There is ipsilateral retention of pain and temperature in lateral hemisection of the cord.

45 *In disease of the spinal cord and CNS:*

a Tabes dorsalis (a manifestation of syphilis) is associated with degeneration of the dorsal root ganglion cell and its processes.

b The herpes virus often lies quiescent in axons for years.

c Hyperacusis can occur in Bell's palsy.

d Taste from the anterior two-thirds of the tongue is affected in a supra-nuclear VIIth nerve palsy.

e Acute demyelination of spinal roots/nerves can cause glove-and-stocking sensory symptoms.

Tracts in the cord (Answers)

42 a **True.** Also the lateral corticospinal tract is the major crossed volitional motor tract of the cord.

b **False.** The vestibulospinal tract is uncrossed. It synapses on ipsilateral motorneurons and internuncials.

c **False.** The vestibulospinal tract predominantly stimulates extensor motorneurons.

d **True.** Reticulospinal fibres are concerned with motor function. They are mostly ipsilateral.

e **True.** Vestibulo- and reticulospinal tracts modulate voluntary movement and mediate control of unconscious movement.

43 a **True.** The dorsal columns also carry proprioceptive information (i.e. vibration and position).

b **False.** The lateral spinothalamic tract carries temperature and pain.

c **False.** The tract of Lissauer contains the central processes of first-order sensory neurons which travel for a few segments both up and down the cord.

d **True.** The spinocerebellar tracts also carry impulses from muscle spindles and other proprioceptors.

e **False.** Uncrossed afferent tracts include the posterior columns and the spinocerebellar tracts.

44 a **False.** A lesion above C3 can cause respiratory arrest as well as a quadraplegia.

b **True.** α-Receptors innervate trigone, bladder neck, and internal urethral sphincter.

c **False.** Autonomic reflexes are the first to return after spinal shock.

d **False.** Whiplash is almost certainly the result of an acute hyperextension injury to the neck. It may be caused by overstretching of cervical roots. It typically comes on after a time lapse (hours–days–weeks).

e **True.** Pain and temperature are retained on the side of the lesion because fibres that carry these sensations have already crossed to the other side of the cord.

45 a **False.** In tabes dorsalis the ganglion cells are remarkably unaffected. Degeneration occurs in the dorsal root proximal to the ganglion cell, especially in dorsal and lumbosacral regions.

b **False.** The herpes virus may lie dormant in sensory ganglia (and epithelial cells) for long periods. It is reactivated by a variety of stresses.

c **True.** Hyperacusis will occur if the nerve to the stapedius is involved.

d **False.** Loss of taste implies that the facial nerve is damaged proximal to where it gives off the chorda tympani. The chorda tympani does not travel with the 'extracranial' facial nerve.

e **True.** A glove-and-stocking sensory deficit occurs in the Guillain–Barré syndrome and in some genetic and metabolic deficiency state poly-neuropathies. Lead levels should be checked if a motor neuropathy is present.

Cerebral blood flow (Qs 46–50)

46 *Cerebral blood flow (CBF):*

   a is subject to autoregulation over a mean arterial pressure (MAP) range of 50–150 mmHg.

   b is normally about about 50–55 ml/g/min.

   c can fall to 20 ml/100 g/min without changes in the EEG.

   d is critical for maintaining the intracranial pressure (ICP) at around a normal of 10–15 mmHg.

   e is adjusted to keep a cerebral perfusion pressure (CPP) of about 70 mmHg in normal people.

47 *Regarding cerebral blood flow (CBF):*

   a The cerebral perfusion pressure (CPP) should not be allowed to fall below 30–40 mmHg.

   b Cerebral ischaemia may occur in hypertensives when the MAP falls to 80 mmHg.

   c Intracranial contents can increase in volume by about 10 ml before intracranial pressure (ICP) rises inordinately.

   d A doubling of the normal $PaCO_2$ causes a virtual doubling of CBF.

   e Hypoxia is a more potent stimulus for increased CBF than a similar fall in the $PaCO_2$.

48 *Cerebral blood flow:*

   a Overall CBF and oxygen consumption in the brain varies widely throughout a normal 24 hours.

   b Cerebral ischaemia occurs if CBF delivers less than 5 ml of oxygen/100 g brain tissue/min.

   c Sympathetic vasoconstriction of cerebral vessels is often seen in acute brain damage.

   d Cerebral steal means that blood is diverted to the brain from extracranial regions.

   e Inverse steal means that blood is diverted to good areas of the brain from damaged ones, so that damage limitation is achieved

49 *Factors affecting the CBF and the intracranial pressure (ICP):*

   a The cerebral blood flow increases during sleep.

   b The intracranial pressure increases during sleep (irrespective of position).

   c Halothane selectively increases the cortical blood flow.

   d Almost all volatile anaesthetics increase the intracranial pressure.

   e Colloid oncotic pressure (COP) is a strong driving force for shifting fluid to and from tissue spaces in normal brain.

50 *The following are neuroprotective:*

   a Hypothermia is a powerful agent by reducing cerebral O_2 demand.

   b Inhibition of the breakdown of arachidonic acid (AA).

   c Glutamate inhibitors, e.g. NMDA antagonists.

   d Elevated serum glucose.

   e Induced hypertension.

Cerebral blood flow (Answers)

46 a **True.** Autoregulation is lost outside a range of 50–150 mmHg. MAP < 50 mmHg causes coma; MAP > 150 mmHg causes cerebral oedema.

    b **False.** Normally the CBF is about 50 ml/100 g/min (grey matter, 100 ml/100 g/min; white matter, 20 ml/100 g/min).

    c **True.** Electrical function resists hypoxia better than metabolic function.

    d **True.** ICP is an immediate determinant of cerebral perfusion.

    e **True.** CPP (70) = MAP (80) – ICP (10).

47 a **True.** A CPP less than 30–40 mmHg causes functional brain damage.

    b **True.** Impaired autoregulation occurs in hypertensives, in the elderly, and at all ages after cerebral trauma.

    c **False.** Any rise in volume above 5 ml causes a steep rise in ICP. Total intracranial volume = volume of blood + CSF + volume of brain parenchyma (ICF + ECF).

    d **True.** Likewise, a fall in the $PaCO_2$ causes a steep fall in CBF.

    e **False.** The PaO_2 must fall below 50 mmHg (6.6 kPa) before there is a significant rise in the CBF.

48 a **False.** CBF and O_2 consumption of brain (3 ml/100 g/min) are fairly constant over a 24 h period.

    b **False.** The CBF should deliver over 2 ml of oxygen per 100 g brain tissue/min, Otherwise immediate or delayed neuronal death will occur.

    c **True.** Sympathetic vasoconstriction may occur in the larger cerebral vessels after a subarachnoid bleed. Calcium blockers may be helpful in such cases. Maintaining a relative hypertension is more important. β-Blockers are potentially dangerous. Significant vasoconstriction can be caused by very high levels of amines, e.g. after a haemorrhage elsewhere.

    d **False.** Steal implies the diversion of blood to an area of the brain that is relatively vasodilated from other normally perfused areas of the brain.

    e **False.** The opposite is the case, e.g. deliberate hypocarbia constricts vessels in the normal brain and so diverts blood to ischaemic areas where vessels are unresponsive. This is called the Robin Hood phenomenon.

49 a **True.** CBF increases especially during rapid eye movement sleep.

    b **True.** ICP rises in sleep and especially if there is an intracranial mass, when it can rise from a normal 15 mmHg up to 50 mmHg.

c **True.** Halothane produces a 20–40% decrease in cerebral vascular resistance in normocapnic people with a minimum alveolar concentration (MAC) of 1.2–1.5%. Yet it decreases subcortical blood flow.

d **True.** In equivalent MAC doses nitrous oxide is probably a more powerful cerebral vasodilator than either halothane or isoflurane.

e **False.** For example, a reduction of COP with maintenance of serum osmolarity after IV normal saline increases the water content of many tissues (where pore size is about 65 Å) but not of the brain (tight junctions, 7 Å wide). Nor does ICF water in brain decrease significantly after infusion of a hypertonic electrolyte solution.

50 a **True.** There is a 50% reduction in O_2 demand as temperature falls from 37 to 27 °C.

b **False.** FFAs increase during ischaemia. On reoxygenation arachidonic acid is broken down to damaging PGs and leukotrienes.

c **True.** This is because glutamate and similar excitotoxins are released from damaged neurons.

d **False.** Elevated plasma glucose is damaging. It causes a greater than expected fall in intracellular pH. Plentiful glucose provides excess lactic acid during ischaemia. Furthermore, glucose-induced arteriolar dilation impairs effective vasoregulation and facilitates hypertensive damage to all organs.

e **True.** This is controlled 'triple H' therapy: hypertension to increase perfusion, haemodilution to improve rheology; and hypervolaemia to do both.

Cerebral cortex, EEG, and CVA (Qs 51–57)

51 *In the cerebral cortex:*
a the grey matter is normally about 1 cm thick.
b Granule (stellate) cells are plentiful in the primary sensory cortex.
c An understanding of concepts is predominantly a function of the right hemisphere.
d The bands of Baillarger are well developed in the sensory areas.
e Over 90% of corticospinal fibres are myelinated.

52 *Cortical speech centres:*
a Both sides of the brain are needed for sensible fluent speech.
b In left-handed people Broca's and Wernicke's areas are mostly found in the right cerebral hemisphere.
c Wernicke's and Broca's areas are situated in the same gyrus.
d In damage to Wernicke's area (sensory aphasia) comprehension of written language is retained.

e   Destruction of Broca's area (motor aphasia) causes complete loss of speech while comprehension is retained.

53 The electroencephalogram (EEG):

a   The EEG is due to firing of cortical neurons (AP generation).
b   In an alert adult with closed eyes, alpha waves predominate.
c   Faster frequencies tend to predominate in most diseases that affect cerebral function.
d   Irregular small waves are characteristic of mental activity.
e   Visual evoked potentials (VEPs) normally peak at about 100 ms.

54 In epilepsy:

a   Seizures result from paroxysmal asynchronous discharge of cortical neurons.
b   Very high spikes occur during the seizure.
c   During seizures the brain ECF K^+ often rises dramatically.
d   Most patients eventually progress to an intractable stage.
e   Virtually all antiepileptic drugs work by increasing the GABA:glutamate ratio.

55 In cerebrovascular accidents (CVA):

a   Most cases are due to haemorrhage.
b   An embolus of only 1 mm across can block any vessel on the cerebral surface and trigger a stroke.
c   Prothrombotic states include a rise in antithrombin III.
d   A transient ischaemic attack (TIA) usually lasts < 10 min.
e   A lacunar stroke causes non-focal damage (i.e. no seizures, no specific motor or sensory deficit).

56 In the natural progression of cerebrovascular accident (CVA):

a   Hyperreflexia is typical in a completed CVA.
b   Hypertonia is especially marked in a pure 'pyramidal' lesion, e.g. in the medullary pyramid.
c   Excellent recovery can be expected after a pure pyramidal lesion.
d   Aspirin definitely reduces incidence of stroke after a TIA.
e   Extracranial–intracranial bypass is very useful in preventing stroke where there is carotid artery stenosis.

57 The thalamus:

a   plays an essential role in arousal.
b   is important for cognition and awareness.
c   acts as an important sensory relay station.
d   has each half of the body represented topographically.
e   after infarcting, can have spontaneous excessive pain.

Cerebral cortex, EEG, and CVA (Answers)

51 a **False.** In most people, right-sided input is not needed for sensible speech. However, pure left-side speech is without animation or emotional significance.

   b **True.** In contrast, pyramidal cells (Betz) cells are plentiful in the motor precentral gyrus.

   c **True.** In general, the right side is the 'thinker', the left side the 'doer'.

   d **True.** The bands of Baillarger are mainly collateral and terminal branches of afferent fibres (thalamocortical fibres). They run parallel to the cortical surface.

   e **False.** 20–30% of the fibres in the corticospinal tracts are unmyelinated C fibres. Their origin and function is unclear.

52 a **False.** The speech centre is bilateral although predominantly left-sided. Pure left-sided speech is without animation or emotional significance.

   b **False.** Speech centres are found in the right hemisphere or in both in only 15% of left-handers.

   c **False.** Note: Wernicke's area is in posterior superior temporal lobe; Broca's is on the inferior frontal gyrus.

   d **False.** Sensory aphasia is where comprehension of spoken and sometimes written language is lost.

   e **False.** Motor aphasia is the loss of spontaneous speech while comprehension is retained. Short, halting, telegraphic sentences can be formulated.

53 a **False.** The EEG is due to changes in EPSPs and IPSPs in cortical neurons.

   b **True.** Basic alpha rhythm is 8–10/s, 50 μV in amplitude.

   c **False.** Theta and delta waves are 3–6/s, 80–120 μV and mostly occur in sleep and in diseases that affect cerebral function.

   d **True.** Mental activity is associated with irregular beta waves at a rate > 13/s. Local functional activation causes an increase in local blood flow that parallels glucose usage but may be far in excess of oxygen consumption.

   e **True.** VEPs are delayed especially in demyelinating diseases, e.g. multiple sclerosis.

54 a **False.** Epilepsy is characterized by abnormal synchronous neural discharge.

   b **True.** Spikes during a seizure are often 1000 μV high. However, high spikes are not diagnostic of epilepsy.

   c **True.** During a seizure the ECF K^+ in brain may rise from 3 to 10 mmol/l but ECF Ca^{2+} and Na^+ decrease.

   d **False.** 20–30% of cases of epilepsy become intractable. Many doctors hold that early and continued anticonvulsant treatment prevents intractable epilepsy.

e **False.** Valproate and benzodiazepines enhance GABA. Ethosuximide and anti-NMDA rectifier dugs inhibit spike-generating Ca^{2+} currents in thalamus and elsewhere; phenytoin and carbamazepine stabilize Na^+ channels.

55 a **False.** About 80% of CVAs are due to infarction; about 20% are due to haemorrhage.

b **True.** Furthermore, intracardiac thrombi < 4–8 mm which may cause cerebral emboli are not visualized by current methods.

c **False.** Prothrombotic states include a fall in antithrombin III, protein C, protein S, and a rise in anticardiolipin antibodies.

d **True.** Occasionally a TIA lasts up to 24 h, after which it is a CVA by definition. The most common cause is an atheromatous narrowing of the extracranial carotid artery.

e **True.** A lacunar stroke is due to vascular injury or disease deep in noncortical parts of brain or brain stem. It accounts for 20% of CVAs.

56 a **True.** A completed CVA can have all the features of an upper motor neuron lesion.

b **False.** A pure 'pyramidal' lesion leaves the powerful vestibulospinal and reticulospinal tracts intact.

c **True.** Many apparently voluntary acts come back despite some clumsiness and weakness after a pure 'pyramidal' lesion.

d **True.** Aspirin reduces the risk of stroke after a TIA by 25–30%. Note that the risk of stroke after a TIA is 6–7% per year.

e **False.** This was proved false by the extracranial–intracranial bypass study (1985).

57 a **False.** Arousal is possible even if the thalamus is grossly damaged.

b **True.** The disproportionate damage to the thalamus compared to the cerebral cortex in the well-known case of Karen Ann Quinlan (1994) supports this view.

c **True.** Ascending sensory fibres synapse in the thalamus and are projected to all parts of the cerebral cortex.

d **True.** They are represented topographically in the contralateral thalamus.

e **True.** This is the thalamic syndrome. Pain is spontaneous or precipitated by minor stimuli. Paradoxically intractable pain may be successfully treated by thalamic lesions induced surgically.

Subcortical areas (Qs 58–63)

58 In the cerebellum:

a The axons of the Purkinje cells constitute the sole output of the cerebellar cortex.

b Each hemisphere connects with the opposite cerebral cortex.

c There is no sensory loss, motor weakness, or mental impairment in pure cerebellar disease.

d A lesion of the vermis typically causes dyscoordination of the limbs.

e Ataxia is worsened when the eyes are closed.

59 Regarding the reticular formation (RF) and the limbic system:

a The RF is a loose collection of neurons and fibres extending through the axis of the CNS from spinal cord to cerebrum.

b The only proven functions of the RF are associated with wakefulness, arousal, and posture control.

c The hypothalamus is the main outlet for the limbic system.

d The amygdaloid nucleus is mainly concerned with memory.

e The hippocampus is important in behavioural disorders.

60 In the basal ganglia:

a some striatonigral fibres release acetylcholine.

b some striatonigral fibres release serotonin.

c some nigrostriatal fibres secrete dopamine.

d some nigrostriatal fibres secrete acetylcholine.

e the caudate nucleus and the putamen constitute the neostriatum.

61 Idiopathic Parkinson's disease (IPD):

a For clinical symptoms to arise the dopamine content of the striatum must be reduced by at least 50%.

b The nigrostriatal pathway is the most severely affected in Parkinson's disease.

c D_2-receptors are increased in the striatum of untreated patients.

d Dopamine inhibits cholinergic interneurons and stimulates GABAergic ones in the neostriatum.

e Decreased nigral GABA is found in many cases of Parkinson's disease.

62 Parkinson's disease (PD) and parkinsonism:

a Methyl-phenyl tetrahydropyridine (MPTP) induced parkinsonism is a progressive condition once it is established.

b The tremor in PD is exacerbated by alcohol.

c The tremor of PD is present even when the subject is asleep.

d The 'on–off' phenomenon usually occurs in the first year of L-dopa treatment.

e GABA agonists worsen parkinsonism.

63 In idiopathic Parkinson's disease:

a Anticholinergic drugs are very helpful in the control of hypokinesis.

b Inhibitors of monoamine oxidase B are useful in the treatment of PD.

c A deficiency of caeruloplasmin may present with parkinsonism.

d Full expression of tardive dyskinesia (TD) requires an intact dopaminergic nigrostriatal pathway.

e Ingestion of fat exacerbates the 'on–off' phenomenon.

Subcortical areas (Answers)

58 a **True.** The axons of the Purkinje cells are GABAergic and inhibit the deep cerebellar nuclei, e.g. the dentate nucleus.

b **True.** Since both afferent and efferent cerebellar fibres cross, each cerebellar hemisphere controls its own side of the cord.

c **True.** The cerebellum controls co-ordination and force of goal-directed movement.

d **False.** The unpaired vermis controls mid-line co-ordination in the head and trunk. It is often involved in medulloblastoma in children.

e **False.** Visual input matters little since motor activity is still misguided by the damaged cerebellum. This contrasts with the positive Romberg test in posterior column disease.

59 a **True.** The RF has exceptionally long dendrites and axons.

b **False.** The RF also modulates muscle tone, maintains vegetative functions, and has a role in affective emotions (aggression, pain, pleasure).

c **True.** However, most parts of the RF, such as the hippocampus, connect also to the thalamus, mid-brain, and further afield.

d **False.** The amygdaloid nucleus is important for aggression, restlessness, and instinctual drives.

e **False.** The hippocampus appears to be important in short-term memory.

60 a **False.** Striatonigral fibres are GABAergic and depress dopamine neurons in the substantia nigra.

b **False.** Some nigrostriatal fibres release 5-HT. This may have a similar function to dopamine.

c **True.** Dopaminergic nigrostriatal fibres inhibit muscle tone and improve muscle kinesia.

d **True.** Cholinergic nigrostriatal fibres excite tone via basal ganglia internuncials. They are inhibited by DOPA.

e **True.** The globus pallidus is the palaeostriatum.

61 a **False.** Striatal dopamine must be reduced by at least 70–80% before the clinical symptoms of PD arise.

b **True.** In PD there is cell loss especially in the pars compacta of the substantia nigra.

c **True.** An increase in the D_2-receptors is possibly receptor up-regulation in compensation for the decrease of local dopamine.

   d **False.** Dopamine inhibits both cholinergic and GABAergic fibres.

   e **True.** Total nigral GABA may fall in PD because of loss of its target cells (the DOPA neurons of the substantia nigra).

62 a **False.** MPTP-induced parkinsonism is not progressive. Idiopathic Parkinson's disease is progressive.

   b **False.** The tremor of PD is lessened by alcohol. It is a coarse, pill-rolling tremor usually at a rate of 4–5 Hz.

   c **False.** The tremor of PD disappears in sleep. It is also lessened by voluntary movement.

   d **False.** The 'on–off' phenomenon usually occurs after 5–10 years of L-dopa control. It is due to a too rapid onset and termination of therapeutic effects and increased variation in response to L-dopa.

   e **True.** GABA agonists, such as baclofen, sodium valproate, and hydantoin, worsen parkinsonism.

63 a **False.** Anticholinergic drugs have useful effects on rigidity and especially on tremor. Benzhexol and orphenadrine are most often used.

   b **True.** An example is selegiline. MAO B degrades DOPA but does not degrade NA or 5-HT. COMT inhibitors may help in PD too.

   c **True.** This occurs as in Wilson's disease, i.e. hepatolenticular degeneration.

   d **True.** Tardive dyskinesia is probably due to supersensitivity of brain DOPA receptors, or to an excess of cerebral dopamine.

   e **False.** Ingestion of protein aggravates the 'on–off' phenomenon. Amino acids interfere with the Na^+-dependent transport of L-dopa across membranes.

Pain (Qs 64–65)

64 In the transmission of painful stimuli:

   a Pain receptors are polymodal in nature.

   b Sharp, rapid, intense pain is carried by heavily myelinated fibres.

   c Unmyelinated C fibres only carry slow, dull, diffuse, aching pain.

   d Some pain fibres remain uncrossed in the spinal cord.

   e The 11-amino-acid peptide substance P is mostly released at pain receptors.

65 In the neural modulation of pain:

   a Raphespinal tracts are serotoninergic.

   b Large-diameter sensory neurons depress activity in pain fibres by releasing ACh in the dorsal horn.

   c Anaesthetization of a stump gives only temporary relief in phantom pain.

   d Pain pathways are not fully developed in the newborn.

   e Itching is subserved by a specific set of neurons.

Pain (Answers)

64 a **True.** Pain receptors respond in the same way to a variety of dissimilar stimuli.

   b **False.** Sharp pain is carried by A δ finely medullated fibres at 10–30 m/s. These terminate mostly in lamina V of the dorsal horn, the nucleus proprius.

   c **False.** C fibres carry many sensations at 2 m/s. Some carry only pain and itch.

   d **True.** The palaeospinothalamic tract synapses in ipsilateral raphe nuclei.

   e **False.** Substance P is mostly in the dorsal root ganglion cells and is secreted at central primary afferent terminals. (see Ch 7 Q3a)

65 a **True.** The raphespinal tracts may also release GABA, NA, DOPA, opioids, and somatostatin.

   b **False.** Large-diameter sensory neurons depress activity in pain neurons by releasing GABA or glycine. For more information see the gate control theory of Melzack and Wall (1965).

   c **False.** Phantom pain is central in origin. Anaesthetization of an amputation stump gives no relief.

   d **False.** Pain receptors have been demonstrated from the 7th week in the fetus. Thalamocortical connections are complete by 20–24 weeks of gestation.

   e **True.** Itch has a similar but separate pathway from pain. It is unaffected by analgesics. Pruritus is found in uraemia, obstructive jaundice, Hodgkin's disease, polycythemia, and diabetes. Treatment includes cholestyramine, antihistamines, erythropoietin, and sedation.

Memory (Q. 66)

66 Learning and memory:

   a Short-term memory (primary memory) has a small capacity.

   b Working memory (similar to short-term memory) is located mainly in the frontal lobes.

   c The long-term potentiation of neuronal signals in memory depends mainly on 5-HT.

   d One cannot access short-term memory in retrograde amnesia.

   e One cannot transfer information from primary to secondary memory in anterograde amnesia.

Memory (Answer)

66 a **True.** Short-term memory has a small capacity, is rapidly accessed, and lasts only a few seconds.

b **True.** Dopamine is an important NT in working memory.

c **False.** NO/glutamic acid, PGs, and PAF (platelet activating factor) are the chief candidates for the long-term potentiation of memory.

d **False.** One cannot access secondary or long-term memory in retrograde amnesia.

e **True.** Anterograde amnesia is often seen in organic dementia (e.g. Korsakoff syndrome and alcoholism). This may be due to hippocampal dysfunction or degeneration of mesolimbic pathways.

Sleep and coma (Qs 67–70)

67 *Rapid eye movement (REM) sleep:*

a In REM sleep there is a general rise in skeletal muscle tone.

b REM generating neurons lie in the pons.

c Normally REM sleep occurs within 10 min of falling asleep.

d At birth REM accounts for about 50% of sleep.

e Nocturnal cardiac asystole is most common during REM sleep.

68 *Pathophysiology of sleep:*

a Anxiety and irritability are associated with deprivation of REM sleep despite adequate NREM sleep.

b Many depressive patients have shortened stage 3–4 NREM sleep.

c Sleep-walking occurs in deep stage 3–4 NREM sleep.

d Hypnic jerks, bruxism (teeth grinding), and head-banging occur in REM sleep.

e Recurrent short episodes of NREM during the day cause little or no trouble for those so affected.

69 *In disorders of sleep:*

a Daytime sleepiness, pre-sleep dreams, sleep paralysis, and cataplexy confirm the diagnosis of narcolepsy.

b During sleep it is abnormal for apneic periods to recur more than five times per hour.

c Central sleep apnea is accompanied by exaggerated respiratory movements.

d Obstructive sleep apnea and obesity often go together.

e Polycythemia is common in subjects who suffer from sleep apnea.

70 *In depressed consciousness:*

a The Glasgow Coma Scale is an objective assessment of the severity of coma.

b The sleep–wake cycle is retained in coma.

c Vegetative states are accompanied by purposive movements.

d Speech is lost in the locked-in syndrome.

e Brain death is assumed if the EEG is isoelectric for 30 min.

Sleep and coma (Answers)

67 a **False.** In REM sleep atonia is typical except in the ocular muscles.

    b **True.** The pontine REM centre is near the neurons that generate saccadic eye movements.

    c **False.** Normally REM begins about 40 min after sleep onset.

    d **True.** REM sleep falls to about 20% of total sleep in the adult. Its percentage rises again in old age.

    e **True.** REM may also trigger cardiac arrhythmias, night terrors, and cluster headaches.

68 a **True.** Drugs that selectively depress REM give unrefreshing sleep (such as barbiturates, mandrax, chloral hydrate, or alcohol).

    b **True.** Many depressive patients also have shortened REM latency. They enter REM shortly after onset of sleep. Drugs that increase REM, e.g. reserpine, may cause depression. Many antidepressants suppress REM.

    c **True.** Sleep-walkers function at a low level of critical awareness.

    d **False.** All these phenomena occur in stage 1–2 of NREM sleep.

    e **False.** Multiple, recurrent, inappropriate daytime 'microsleeps' leave subjects heedless, forgetful, and unable to concentrate.

69 a **True.** Narcoleptics have inappropriate recurrent episodes of REM sleep.

    b **True.** It is also abnormal for a sleep apneic period to exceed 10 s.

    c **False.** In central sleep apnea there are no respiratory efforts despite the rise in $PaCO_2$ and the fall in the PaO_2.

    d **True.** Obstructive sleep apnea and obesity go together (for example the fat boy in the *Pickwick Papers*).

    e **True.** Polycythemia is common because of repeated hypoxic episodes.

70 a **True.** The Glasgow Coma Scale is based on stimulus for eye-opening and best motor and verbal responses. A score of less than 10 out of 15 indicates serious damage. A dead person will even score three!

    b **False.** However, the sleep–wake cycle is retained in persistent vegetative states.

    c **False.** Non-purposive chewing and grimacing occur in vegetative states.

    d **True.** In the locked-in syndrome communication is by eye movements or blinking only.

    e **False.** The following may be used as tests of brain death: an isoelectric EEG for 60 min (used in the USA); fixed and unresponsive pupils; absence of corneal, vestibulo-ocular, respiratory, and brain stem reflexes. One should also exclude any possibility of reversible coma from metabolic disorders.

Temperature and disorders (Qs 71–83)

71 Body temperature:

a It is normally highest around 18.00 h and lowest around 06.00 h.

b It remains the same whether one is in the Arctic or in the tropics.

c Temperature gradients exist between extensor and flexor surfaces.

d Body location of some diseases is related to local temperature.

e Surface thermoreceptors fire steadily even at a comfortable ambient temperature.

72 Thermoperception:

a The end organs of Krause and Ruffini subserve cold and heat.

b Thermoreceptors that respond to cold (cold spots) are more plentiful than those that respond to heat (hot spots).

c Maximal discharge from 'hot spots' is between 40 and 45 °C.

d Cold spots show a rapid discharge when the temperature exceeds 45 °C.

e False thermoperception often occurs in normal people.

73 Concerning thermoregulation:

a Radiation is the most important source of our body heat.

b Radiation is the major route by which the body loses heat.

c Water immersion reduces heat loss by convection.

d Water immersion increases heat lost by conduction.

e Thermal conductivity of the body as a whole is poor.

74 In the response to cooling:

a Cutaneous vasoconstriction occurs until skin temperature falls below 0 °C.

b Shivering can occur when one is asleep.

c Shivering is a relatively inefficient way of generating body heat.

d thyroid hormone increases significantly.

e All the 'adult' thermoregulatory responses occur in the newborn.

75 In hypothermia:

a The pH decreases as the body temperature is lowered.

b The PR interval is increased in the ECG.

c Physical activity always slows the onset of severe hypothermia.

d Survival is shorter in cold immersion if one's clothes are sodden than if one is naked.

e Adults survive cold exposure better than small children.

76 Thermoregulation:

a Differing heat loss mechanisms occur in response to endogenous- and exogenous-mediated hyperthermia.

b   Vasodilation is of no help if the ambient temperature exceeds that of the body.
c   Minimal basal sweating is normal at rest in temperate climates.
d   Thermal sweating is largely abolished by sympathectomy.
e   Eccrine glands are more sensitive in heat-acclimatized people.

77 *In sweating:*

a   Sweat is close to isotonic at high rates of secretion.
b   A hot, humid, breezeless environment reduces the secretion of sweat.
c   Thermal sweating is impaired by general anaesthetics.
d   Extensive skin burns increase sweat fluid loss.
e   Non-thermal sweating occurs via the apocrine glands.

78 *About hyperthermia:*

a   In moderate exercise there is an upward resetting of the hypothalamic thermostat.
b   In exertional heat stroke the arterial lactate seldom exceeds 5 mmol/l.
c   One can survive longer in moist than in dry heat.
d   Heat stroke comes on once the core temperature reaches 50 °C.
e   Body temperature does not rise when taking a comfortable hot bath.

79 *Hyperthermia:*

a   Patients with thyrotoxicosis usually have a normal temperature.
b   Patients with phaeochromocytoma usually have a normal temperature.
c   Dehydration is a common cause of hyperthermia.
d   In the neuroleptic malignant syndrome there is excess sensitivity to ACh.
e   In therapeutic hyperthermia the body temperature is raised to as high as 45 °C for several hours.

80 *Pathogenic fever:*

a   This is caused by an up-regulation of the thermoregulatory set point in the preoptic nucleus of the hypothalamus.
b   Exogenous pyrogens release prostaglandin (PG) mediators in the hypothalamus.
c   The major endogenous pyrogen is a thermogenic amine.
d   Endogenous pyrogens increase hypothalamic PG of the E series.
e   Endogenous pyrogens inhibit secretion of corticotrophin-releasing factor (CRF).

81 *In fever:*

a   Cortisol suppresses fever mainly by inhibiting PG synthesis.
b   AVP is a very strong candidate as an endogenous febriolytic.
c   Resultant elevated temperatures are useful in disease.
d   The febrile response is useful in stress states.
e   Neurogenic fever, e.g. from cranial tumour or head injury, is mediated by prostaglandins.

82 *General anaesthesia and thermoregulation:*

a Malignant hyperthermia (MH) is an autosomal dominant condition.

b Core hypothermia is a prerequisite for postoperative shivering.

c General anaesthesia (GA) increases activation threshold for responses to hypothermia (viz. sensitizes them).

d GA increases the activation response threshold against hyperthermia (viz. desensitizes them).

e Thermoreceptor silence is prolonged in GA (see answer 71e).

83 *Effects of general anaesthesia on temperature:*

a The vasoconstriction threshold in infants on halothane or isoflurane is far lower than that for adults.

b The thermoregulatory threshold (TRT) for propofol/N_2O is almost normal.

c Core temperature decreases rapidly during the first hour of GA.

d Initial hypothermia is due to heat loss exceeding heat production.

e Almost 50% of metabolic heat is lost via respiration when the patient is ventilated with dry, cool gas.

Temperature and disorders (Answers)

71 a **True.** Body core temperature swing is about 1 °C and is most marked in women and children.

b **False.** Body temperature rises a little in hot climates and falls a little in cold ones. This flexibility reduces work expended on homeothermy.

c **True.** The most marked temperature gradient is between body core and shell.

d **True.** Examples include pediculosis at flexor creases, leprosy on extensor surfaces, and kala-azar in the internal viscera.

e **False.** An ambient range of 30–40 °C (thermal neutrality) is associated with virtual 'thermoreceptor' silence in receptor adaptation.

72 a **False.** Krause and Ruffini organs subserve vibration and pressure, i.e. sensation.

b **True.** Cold spots are mostly in superficial layers of skin, especially in fingers and toes. Hot spots lie deeper.

c **True.** Cold spots are virtually silent at between 40 and 45 °C. Pain occurs above 45 °C (and below 15 °C).

d **True.** Any discharge from cold spots above 45 °C is interpreted as cold, hence paradoxical cold response. Pain comes from tissue damage caused by heat acting on free nerve endings.

e **True.** For example, cold is not felt by warm skin, e.g. rolling in snow after a sauna. Cold can be felt in a warm environment, e.g. as a hot bath cools, one may feel cold although the water is still warmer than the body.

73 a **False.** Endogenous thermogenesis is most important. Radiation is usually the most important external source of heat.

b **True.** Radiation accounts for 50% of heat loss (unclothed), evaporation of water from skin accounts for 30%; convection 10%; conduction 5%; and losses from respiratory system, etc. 5%.

c **False.** Specific heat capacity of water is 1000 times greater than that of air. Note there are also fluid currents.

d **True.** Heat conductance of water is 25 times that of air.

e **True.** Vascular transfer of heat from core to shell is vital. Vasodilation increases the thermal conductivity of fat tenfold.

74 a **False.** Vasodilation occurs between 5 and 10 °C. There is altered response of cold receptors and vasoparalysis below 0 °C.

b **True.** This emphasizes involuntary nature of shivering.

c **False.** Shivering rapidly increases heat generation up to fivefold and more.

d **False.** In humans, cold only provokes a small increase in TRH.

e **True.** The newborn depends also on the metabolism of brown fat.

75 a **False.** The pH increases at low temperatures (due to a change in the ionization constant of water, K_w).

b **True.** There is delayed conduction of the cardiac impulse in hypothermia. ST deviation and J waves may also occur.

c **False** Exercise increases the rate of heat loss. Exercise is obviously useful in short-term cold exposure.

d **False.** In freezing water one survives 10 min naked, 20–30 min in sodden clothes, and several hours in waterproofs.

e **True.** A child's core temperature must fall to <25 °C (<30 °C in adult) before cardiac standstill. Rapid cooling of a child causes a rapid decrease in O_2 requirements (to one-quarter at 25 °C). Because their surface area to volume ratio is much greater, a small child cools much more rapidly than an adult and so will die sooner on exposure to low temperatures.

76 a **False.** The same heat loss responses occur irrespective of the source of the hyperthermia.

b **False.** Vasodilation creates an insulating fluid envelope and transfers heat to non-vital subcutaneous tissues.

c **False.** Sweating commences abruptly at a skin temperature of 30–31 °C (just above the thermoneutral zone) or if the core temperature rises more than 1 °C.

d **True.** Thermal sweating is also abolished by selective damage to the hypothalamus.

e **True.** Also, duct sensitivity to aldosterone is also increased in heat-acclimatized people.

77 a **True.** However, at slower rates of secretion the sweat ducts have time to extract NaCl and secrete K^+ so that sweat becomes hypotonic.

b **False.** Hot, humid conditions reduce the evaporation and not the production of sweat.

c **True.** Thermal sweating is impaired by general anaesthetics, atropine, phenothiazines, and MAO inhibitors

d **False.** Burns destroy sweat glands and decrease the possible sweat volume.

e **False.** Non-thermal sweating is mainly palm–sole sweating but may involve all eccrine glands.

78 a **True.** In moderate exercise there may be a release of endogenous pyrogen from exercising muscles.

b **False.** In exertional heat stroke the arterial lactate may rise to 10–20 mmol/l and yet survival is possible.

c **False.** This is because sweat evaporates more rapidly in dry heat. Transpirational loss is also greater, leading to dehydration, and heat stroke occurs sooner than in an arid than in a humid environment with similar ambient temperatures.

d **False.** Heat stroke is present once core temperature is 40–41°C. Death occurs at a core of greater than 50 °C.

e **False.** Bath water is usually 43–44 °C. After 30 min, the temperature of the mouth is about 38 °C, heart rate is up to 120/min, and the respiration rate rises to about 28/min.

79 a **True.** Temperature may be normal or only slightly elevated in thyrotoxicosis. This attests to the efficacy of thermal homeostasis. In a thyrotoxic crisis the rectal temperature is >40 °C.

b **False.** Temperature is high in a crisis in phaeochromocytoma. NA vasoconstricts and increases the metabolism.

c **True.** Volume depletion leads to cutaneous vasoconstriction and decreased sweating which impair heat dissipation.

d **False.** The neuroleptic malignant syndrome is probably triggered by blockade of D-receptors in the basal ganglia. Muscle spasticity increases heat production. Also one finds impaired hypothalamic thermoregulation and autonomic dysfunction.

e **False.** A value of 42.4 °C appears tolerable for a few hours, e.g. as an adjunct for cancer therapy.

80 a **True.** Pathogenic fever is maintained by the same physiological mechanisms that are used to maintain normal body temperature in a cold environment.

b **False.** Exogenous pyrogens release endogenous pyrogen from WBCs and tissue macrophages.

c **False.** Cytokines, especially IL-1b and to a lesser extent TNF, IL-6, and interferon, are the major endogenous pyrogens.

d **True.** Endogenous pyrogens pass into the organum vasculosum via loose endothelial junctions. They stimulate local neurons and glia to make PGs, mainly PGE_2.

e **False.** Cytokines affect many hypothalamic nuclei. They stimulate CRF release, inhibit appetite, and induce malaise and somnolence.

81 a **False.** Cortisol inhibits endogenous pyrogen synthesis.

b **True.** AVP, as a febrolytic, comes mainly from extrahypothalamic sites. CRF and MSH are less likely candidates.

c **True.** Fever improves the bactericidal efficacy of macrophages.

d **True.** In stress states there is a cytokine-induced release of acute phase proteins, a switch to lipolysis and proteolysis, and a reduction in divalent cations.

e **True.** Neurogenic fever is mediated probably by PGE via a central IL-1 system. Psychogenic fever is likely to have the same basis.

82 a **True.** MH is triggered by halogenated anaesthetics and depolarizing muscle relaxants. This is screened using the caffeine–halothane contracture test.

b **False.** Postoperative shivering can be abolished without a rise in core temperature by applying radiant heat to face or chest or by giving pethidine (0.3 mg/kg) or doxapram (1.5 mg/kg).

c **False.** GA decreases responses to hypothermia by about 2.5 °C.

d **True.** GA raises the activation threshold for responses to hyperthermia by about 1 °C.

e **True.** During a GA there is a widening of the interthreshold range during which the patient is poikilothermic (i.e. there is thermoreceptor silence).

83 a **False.** The vasoconstriction threshold differs little but it is associated with an increased O_2 consumption from non-shivering thermogenesis in infants.

b **False.** The TRT for propofol/O_2 is 33 °C; for halothane/O_2, enflurane/O_2, and fentanyl/O_2 it is 34.5 °C. Isoflurane decreases TRT by 3 °C multiplied by its % end-tidal concentration.

c **True.** The core temperature falls rapidly in the first hour, followed by slow decrease (2–3 h) then a plateau phase.

d **False.** Initial hypothermia during a GA is mainly due to a redistribution of heat within the body by core to shell shunting.

e **False.** Less than 10% of metabolic heat is lost via respiration. This loss can be prevented by active gas warming and humidification.

Organic basis of CNS disease (Qs 84–86)

84 Neuropharmacology in CNS disease:

a There is an excess of ACh activity in Alzheimer's disease (AD).

b Multiple NT deficits have been found in AD.

c Autism is usually associated with elevated 5-HT in the blood.

d Decreased cholinergic activity in the CNS is often found in depression.

e Selective 5-HT reuptake inhibitors have weaker membrane stabilizing properties than tricyclic antidepressants.

85 *Neurotransmitters:*

a Selective 5-HT uptake inhibitors, e.g. fluoxetine, inhibit liver cytochrome P450 enzyme (CYP2D6).

b Most schizophrenics have decreased levels of dopamine in the CNS.

c D2 blockers, e.g. risperidone, benefit the negative symptoms of schizophrenia.

d Neuroleptic drugs cause extrapyramidal side-effects because they combine with D-receptors in the limbic system.

e Minor tranquillizers, e.g. benzodiazepines, enhance GABAergic transmission.

86 *Changes in the CNS in disease include:*

a The blood–brain barrier is unaffected in multiple sclerosis (MS).

b The inflammatory process in MS is almost certainly immune mediated.

c Primary orthostatic tremor (unsteadiness when standing still) is associated with diminished GABA and/or glycine in the CNS.

d Brain structure appears normal in benign senescence.

e Botulinum toxin is transferred to CNS by axonal transport.

Organic basis of CNS disease (Answers)

84 a **False.** In Alzheimer's disease there is a cholinergic deficit, especially in some subcortical and brain-stem neurons that project to the neocortex.

b **True.** However, replacing ACh deficit of AD by use of an anticholinesterase, e.g. tacrine, seems to give the best results.

c **True.** Yet fenfluramine (decreases 5-HT) is not very helpful. It may be that a neurobiological defect causes autism, e.g. cerebellar vermal hypoplasia.

d **False.** An increased muscarinic activity is more likely in depression. Persons exposed to cholinergic inhibitors often become depressed.

e **True.** Thus selective 5-HT reuptake inhibitors have less quinidine-like effects (are less cardiotoxic) than tricyclics.

85 a **True.** 7% of Caucasians lack CYP2D6 and so are very sensitive to fluoxetine.

b **False.** Many schizophrenics have increased DOPA, increased homovanillic acid (a DOPA metabolite), and increased numbers of D_2-receptors.

c **True.** Perhaps + and–symptoms are related to hyper- and hypo-dopaminergic activity in different parts of the brain.

d **False.** Neuroleptic drugs cause extrapyramidal side-effects because of striatal binding. Clozapine has low striatal binding and has a low incidence of these side-effects.

e **True.** Benzodiazapines bind to a GABA-associated receptor and augment opening of Cl^- channels in hyperpolarization.

86 a **False.** The blood–brain barrier is consistently abnormal in MS, with a focal increase in permeability in areas of inflammation.

b **True.** Immune derangement in MS may be viral linked against an 170-amino-acid myelin basic protein.

c **False.** The cause of primary orthostatic tremor is unknown. High regular frequency muscle activity of 14–18 Hz occurs. This contrasts with Parkinson's disease, with a tremulous frequency of 3–8 Hz.

d **False.** In benign senescence there is a widespread loss of neurons, 'senile plaques' (amyloid deposits), widened sulci, narrowed gyri, and enlarged ventricles.

e **True.** Also in the spinal cord botulinum toxin blocks Renshaw inhibition. In the PNS it inhibits exocytosis of ACh at the motor end plate (see also Q23c).

12

The special senses

Olfaction and taste (Qs 1–2)

1 *Olfaction:*
  a Smell is represented bilaterally in the parahippocampal gyri.
  b Hysterical anosmia can be diagnosed by a sharp response to sniffing NH_3.
  c Receptor organs for smell are free nerve endings of unipolar cells.
  d Only material dissolved in body secretions can be smelled.
  e Anosmia usually follows a fracture of the anterior cranial fossa.

2 *Taste:*
  a Most taste fibres eventually end up in the opposite post-central gyrus.
  b Substances have to be dissolved in order to stimulate taste buds.
  c The sensation of bitterness is mainly detected around the tip of the tongue.
  d The taste receptor cell is a modified nerve cell.
  e Taste buds do not regenerate easily after destruction.

Olfaction and taste (Answers)

1 a **False.** Smell is overwhelmingly and probably uniquely ipsilateral.
   b **False.** A sharp response occurs in organic anosmia because NH_3 stimulates the trigeminal more than the olfactory nerve. Little or no response is found in hysterical anosmia.
   c **False.** The receptor cells in the olfactory epithelium are cell bodies of bipolar neurons. They are remarkable as they are the only neurons that degenerate rapidly. They are replaced constantly by division of other neuroepithelial cells.
   d **True.** Odorants are absorbed into the mucous layer and either combine with a nasal-specific protein or diffuse directly to a protein receptor site. Graded receptor potentials are then triggered via the opening of specific ion channels.
   e **False.** Post-traumatic anosmia only occurs if the fracture involves the cribriform plate.

2 a **True.** Taste fibres eventually end up in the lower end of the contralateral post-central gyrus.
   b **False.** Gustatory cells are chemoreceptors that respond to substances dissolved in oral fluids.
   c **False.** The four primary tastes are: bitter posteriorly, sweet at tip, salt and sour at side of tongue.
   d **False.** Taste receptors are modified epithelial cells. Naked sensory nerve fibres supply these chemosensitive cells.
   e **False.** Taste buds regenerate rapidly. Their numbers decrease significantly after 45 years of age.

Hearing and vestibular function (Qs 3–10)

3 Regarding sound energy:

a Pitch reflects the pressure attained with each sound wave cycle.

b Humans can hear over a range of 40–2000 Hz.

c Loudness is more accurately expressed in bels than in phons.

d The bel scale is logarithmic. Usually expressed in decibels (dBs).

e Absolute lack of sound corresponds with an intensity of zero decibels.

4 Within the ear:

a The tympanic membrane is normally convex into the outer ear canal.

b Flow limitation through the Eustachian tube does not occur in normal people at sea level.

c The ossicles magnify the amplitude of the sound that strikes the tympanic membrane.

d The tensor tympani reacts rapidly to any loud noise (acoustic reflex).

e Vibrations from auditory ossicles are transmitted to the inner ear via the oval window.

5 In the inner ear:

a The perilymph is virtually identical with the CSF.

b The electrolyte composition of endolymph is very similar to ECF.

c The shortest fibres of the basilar membrane are at the apex.

d During sound transmission only individual basilar fibres vibrate.

e There are far more outer hair cells (OHCs) than inner ones (IHCs).

6 In the inner ear:

a At rest inside a hair cell is about –70 mV compared to endolymph.

b Apices of the hair cells face endolymph, and the bases face perilymph.

c The olivocochlear tract (Rasmussen's bundle) goes to both OHCs and IHCs.

d Bending cilia of hair cells in one direction evokes a positive potential and evokes a negative potential (inhibition) if bent in the opposite direction.

e The tunnel of Corti is filled with endolymph.

7 In disorders of hearing:

a A tuning fork on the mid line of the forehead (> 512 Hz) is heard best by the affected ear in middle or outer ear disease (Weber's test).

b A tuning fork (512 Hz) is heard better in front of the ear than when placed on the mastoid process in middle ear disease (Rinne's test).

c A continual 'minivibration' of hair cells is abnormal.

d Ischaemic damage of the basilar membrane can result in tinnitus.

e Tinnitus can be a prominent symptom in polycythemia vera.

8 *In the vestibular apparatus (semicircular canals, utricle, saccule):*

a   The fluid in the vestibular apparatus is separate from that in the scala media.

b   The vestibular nerve originates in bipolar neurons in the internal auditory meatus.

c   Linear acceleration is chiefly sensed by the semicircular canals.

d   A nodding movement of head is detected by the semicircular canals.

e   Small collections of calcium carbonate crystals are often found in the cupola of the semicircular canals.

9 *Regarding vestibular apparatus function:*

a   The semicircular canals respond to all rotational positions of the head.

b   Small changes in the volume of the endolymph cause an illusion of movement which is unrelated to the actual body/head position.

c   A detachment of mineral crystals in the vestibular apparatus is likely to cause vertigo plus tinnitus.

d   Optokinetic nystagmus is typified by a slow involuntary oscillatory eye movement with a fast return.

e   Nystagmus is usually labelled in the direction of slow phase.

10 *In vestibular testing:*

a   The sequence, sitting, then lying down; then turning the head from side to side produces, in normal children, mild to moderate vertigo/nystagmus.

b   As above: a sudden onset of vertigo/nystagmus indicates a central cause for the condition.

c   As above: compound nystagmus (horizontal and rotary) occurs in patients with benign paroxysmal positional vertigo.

d   When lying with head-up tilt of 30 ° the 'horizontal canal' is in a fairly true horizontal plane.

e   Position as in d above: hot water (37–44 °C) in external auditory meatus induces nystagmus in normal people (caloric test).

Hearing and vestibular function (Answers)

3   a   **False.** Pitch depends on the frequency of sound waves. It is expressed in hertz units (1 Hz = 1 cycle/s).

b   **True.** Humans can hear from 20 to 20 000 Hz. This range contracts with age.

c   **False.** 1 phon is the intensity level in decibels of 1000 Hz tone (a sone corresponds to a loudness level of 40 phons).

d   **True.** 1 bel represents a tenfold change in sound energy.

e   **False.** Zero decibel is at the threshold of a normal child's hearing. Some can hear below 0 dB. Conversation occurs at 40–60 db; a factory is 80 db; and a jet plane at 160 db where there is pain and discomfort.

4 a **False.** The tympanic membrane is concave into the outer ear canal, and is elliptical, 1.0 cm high and 0.9 cm wide.

b **True.** There is no flow limitation in the Eustachian tube unless it is narrowed by infection or congestion.

c **False.** The ossicles reduce the amplitude of sound. However, there is a 20-fold increase in the force transmission due to area discrepancies between the footplate of the stapes and the tympanic membrane.

d **False.** Contraction of the stapedius damps down loud noise. The tensor tympani only responds to very loud noise.

e **True.** Vibrations from the oval window travel via the perilymph in the scala vestibuli to the scala tympani and then to the round window.

5 a **True.** Note: the perilymph communicates with the subarachnoid space in young children.

b **False.** Endolymph resembles ICF. It is formed by the stria vascularis.

c **False.** Fibres are longest (membrane widest) at the apex (helicotrema).

d **False.** Usually groups of fibres vibrate together.

e **True.** OHCs are in three rows and there are 20 000 of them. There is only one row of 3500 IHCs.

6 a **False.** Inside a hair cell is −150 mV compared to endolymph; −70 mV compared to perilymph.

b **True.** Very high potentials at rest sensitize hair cells to the slightest movement of the basilar membrane.

c **True.** This cholinergic tract probably dampens hair-cell activity.

d **True.** Thus cochlear hair cells have directional sensitivity.

e **False.** The tunnel of Corti is filled with cortilymph (high in Na^+). Both perilymph and endolymph are toxic to the hair cells.

7 a **True.** The affected ear is believed to be shielded from the effects of room noise.

b **False.** This is because normally air conduction is better than bone conduction.

c **False.** Continual minivibration is a perfectly normal phenomenon. It is not perceived by the cerebral cortex.

d **True.** Ischaemic damage is most common between 50 and 60 years. None the less it can occur much younger patients.

e **True.** Tinnitus in polycythemia may be due to a local thrombosis. Tinnitus also occurs in acoustic schwannoma, after ototoxic drugs (aspirin, quinine, aminoglycosides, loop diuretics), and with local AV aneurysms, depression, Menières disease, or traumatic injury to the ear.

8 a **False.** Endolymph in the vestibular apparatus and the scala media are continuous (via the ductus reuniens).

b **True.** Central fibres travel with the cochlear division of the VIIIth nerve and end in the vestibular nucleus. They relay to the cerebellum, medial longitudinal fasciculus, and vestibulospinal tract (see VOR, Chapter 14 Q. 20).

c **False.** Linear acceleration is chiefly sensed by maculae of the utricles.

d **False.** Cupola cells and endolymph of semicircular canals are of the same density. There must be a mismatch in movement between both for neural discharge. Nodding is chiefly sensed by the saccules.

e **False.** $CaCO_3$ crystals are found in the maculae of the utricle and saccule (otolith organ).

9 a **True.** The semicircular canals also respond to angular movements involving cupola/endolymph mismatch.

b **True.** Vestibular hydrops (raised endolymph pressure), e.g. in Menières disease, causes vertigo.

c **False.** Mineral detachment is most likely only to cause a self-limiting positional vertigo. Minerals are dissolved in time by the endolymph.

d **True.** Optokinetic nystagmus is typified by a slow movement and then a rapid correction. In pendular nystagmus both phases are equal.

e **False.** Nystagmus is always labelled in the direction of the fast phase.

10 a **False.** Occasionally nystagmus may occur in normal elderly people.

b **True.** In benign nystagmus there is a latent period of about 5 s. Symptoms subside within 15 s.

c **True.** This is a positive Hallpike test. In such a case the compound nystagmus is mostly towards the undermost ear, has a long latency, and decays rapidly.

d **False.** The horizontal canal is in the vertical plane when one is lying with a head-up tilt of 30°.

e **True.** In the hot water caloric test the cupola is deflected upward and the ensuing nystagmus (fast phase) is towards the irrigated ear. Cold water does the opposite.

The eye, structure and vision (Qs 11–28)

11 *The cornea:*

a contains very few capillaries.

b is richly supplied with nerve fibres.

c transparency is due to its cells having no nuclei.

d has its outer surface lined by tight-junctioned columnar epithelium.

e develops oedema due to excess corneal tissue fluid.

12 *Regarding the structure of the eye:*

a The sclera is an intact fibrous layer (not pierced by vessels).

b The sclera becomes continuous with the dura mater at the lamina cribrosa.

c The choroid blends gradually into the retina in front and the sclera behind.
d The choroid is highly pigmented.
e The choroid is the chief source of nourishment for the retina.

13 *About the choroid:*

a It is one of the most vascular tissues in the body.
b It helps maintain the intraocular pressure (IOP).
c It has a crucial role in stabilizing intraocular temperature.
d The pupillary margin of the iris is the thinnest part of the iris.
e The iris is well supplied with sensory branches of the trigeminal nerve.

14 *Ciliary body (CB) and lens:*

a The ciliary muscle is inserted into the posterior aspect of the suspensory ligament (zonule) of the lens.
b Ciliary muscle contraction directly reduces zonule elastic tension.
c Numerous sympathetic fibres supply the ciliary muscle.
d The posterior surface of the CB is grooved into about 70 folds (pars plicata).
e The lens grows by the proliferation of central cells.

15 *Pathophysiological changes in of the lens and iris:*

a Age-related sclerosis of lens always causes impaired vision.
b Senile cataract is usually due to abnormal deposition of calcium in the lens.
c Cataractous changes can be due to excessive local hydration of lens.
d Patches of atrophy in the iris suggest glaucoma.
e A tremulous iris when the eyes move rapidly is almost always normal.

16 *In the retina:*

a All layers are derived from neural ectoderm.
b A photoreceptor is inhibited by its specific stimulus.
c A single bipolar cell often synapses with a single cone cell from the fovea.
d Ganglion cells are silent in the dark.
e Usually 'on' ganglion cells are scattered evenly amongst 'off' cells.

17 *Impairment of vision in:*

a albinism can be related to a lack of melanin, e.g in the iris and the pigment layer of the retina.
b melanism can lead to blindness due to excess melanin in the pigment layer.
c familial lipid degeneration, e.g. Tay–Sachs disease, is due to damage of rods and cones.
d Solar retinopathy occurs mostly in the peripheral fields of vision.
e Retinal detachment occurs as the pigment layer and the rest of the retina separate.

18 *The humours in the eye:*

a The vitreous humour is very liable to inflammation in deep, penetrating eye injury.

b   The vitreous humour occupies the posterior chamber of the eye.
c   All the aqueous humour is reabsorbed in the anterior chamber.
d   The total volume of aqueous humour is turned over every hour.
e   Aqueous humour is mainly a passive ultrafiltrate of plasma via the capillaries of the ciliary bodies.

19 Ocular physiology and pathology:
a   In normal conditions the intraocular pressure (IOP) ranges from 10 to 20 mmHg.
b   Ophthalmoscopy is more effective than perimetry in testing for glaucoma.
c   The ophthalmopathy of Grave's disease is mainly due to sympathetic overactivity.
d   The IOP is likely to be increased in the upgaze position in thyrotoxicosis.
e   Contraction of the ciliary muscle lowers the IOP.

20 The iris:
a   Primary open-angle glaucoma is more common in far-sighted people.
b   Elderly people often read better in dim than in bright light.
c   Closed-angle glaucoma is most common in the very elderly.
d   The semi-dilated pupil offers the least resistance to the flow of aqueous.
e   Mydriatic drops may cause prolapse of the iris (iris bombe).

21 In the perception of light:
a   In humans, all the photopigments contain carotinoids derived exclusively from the diet.
b   Apart from the periphery and fovea, the retina has rods and cones in about equal numbers.
c   Cones have a lower threshold to light than rods.
d   The rods react maximally at the yellow/red end of the spectrum.
e   Night blindness is one of the earliest signs of vitamin A deficiency.

22 Regarding rods and cones:
a   Cone pigments regenerate much more slowly than rod pigments.
b   Dark adaptation takes about 5 min to reach maximum.
c   Alcohol dehydrogenase is needed to incorporate vitamin A into the 'visual cycle'.
d   Red-tinted goggles cut out the entry of red wavelengths into the eye and so allow reasonable vision while one is in either a dark or a bright environment.
e   Colour blindness is more common in females than in males.

23 Visual optics:
a   The optical centre or nodal point (N) of the eye is normally close to the fovea.
b   After removal of the lens one can only see near objects.
c   Objects are usually blurred if each eye has marked different refraction.

    d Normal people can discern two light sources 1 mm apart at 10 m.

    e Snellen's types are based on the assumption that the average standard visual angle is 5 minutes.

24 As part of the visual pathways:

    a A complete visual field is represented in each lateral geniculate body.

    b The fovea projects mainly into the contralateral optic tract.

    c Only the central parts of the retina are topographically aligned in the primary visual area.

    d Gross vision can be sensed in the superior corpora quadrigemina.

    e Frontal eye fields control ipsilateral eye movements.

25 Ocular movements:

    a Axes of eyeball are always parallel in normal people (conjugate movement).

    b Saccadic movements occur when eyes follow a moving object.

    c Saccadic movements are generated in the nuclei of third, fourth, and sixth cranial nerves.

    d Visual acuity is heightened during saccadic movements.

    e Lesions that involve the medial longitudinal bundle often cause diplopia.

26 The eye reflexes:

    a Damage to an optic nerve can block the consensual light reflex.

    b The mid-brain Perlia is vital in mediating the light reflex.

    c The optic nerve is an afferent limb of the blink reflex.

    d Normal regular unconscious blinking is central in origin and is not initiated by corneal stimulation.

    e Blepharospasm can be alleviated by interrupting the facial nerve.

27 Pupillary reflexes and reactions:

    a The consensual response has the same magnitude as the direct response.

    b A lesion involving the mid-brain decussation of optic fibres can cause bilateral loss of light reflexes with retention of the near reflex, lid reflexes, and psychosensory reactions.

    c A lesion of the ciliary ganglion can cause an ipsilateral Argyll Robertson pupil.

    d The stimulus to accommodation is a blurring of the actual image.

    e Pupillary dilation due to non-ocular sensory stimuli is solely from stimulation of sympathetic fibres to the pupil.

28 In the control of pupillary size:

    a It is normal to have oscillations of the pupil on exposure to light before it settles to its final size.

    b It is normal for a light-constricted pupil to slowly dilate despite keeping the light source constant.

    c Local irritative lesions of the IIIrd nerve usually affect the sympathetic more than parasympathetic control.

d A progressive unilateral extradural haemorrhage causes ipsilateral pupillary constriction prior to dilation.

e A Horner's syndrome can cause meiosis, ptosis, enophthalmos, corneal hyperaemia, anhydrosis, and increased temperature on the affected side.

The eye, structure and vision (Answers)

11 a **False.** The cornea has no capillaries. It is nourished from those in the cornea/scleral junction, the tear film, by diffusion from the aqueous humour, and from the atmosphere.

b **True.** Nerve terminals in the cornea have an obvious protective function.

c **False.** Transparency of the cornea is due to the regular hexagonal lattice arrangement of its stromal collagen fibres.

d **False.** The cornea is lined by a six-layered stratified squamous epithelium.

e **False.** Corneal oedema is due to an excess hydration of its stromal cells. Oedema is normally prevented by an inner lining of metabolically active cells (Descemet's membrane). There is no ECF in the cornea.

12 a **False.** The sclera is pierced by numerous vessels and nerves which are natural conduits for the spread of disease.

b **True.** The sclera and dura are continuous at the exit point of the optic nerve, medial to the optic axis.

c **False.** The choroid is distinct form both retina and sclera. The lamina vitrea separates it from the retina.

d **False.** The pigment layer is part of the retina. Note: easy visualization of choroidal blood vessels in albinos.

e **True.** Thus irreversible visual damage may follow retinal detachment.

13 a **True.** The choroid is largely made up of a vast network of capillaries.

b **True.** Flow in the choriocapillaries falls if the IOP rises and vice versa.

c **True.** The large choroidal blood flow acts as a heat exchanger.

d **False.** The thinnest part of iris is the root, and rips occur here.

e **True.** However, visual reflexes involving pupil size depend on retinal receptors and not on those in the iris.

14 a **False.** The ciliary muscle arises at the corneoscleral junction and is inserted into the anterior aspect of the ciliary body (CB) and zonule.

b **False.** The ciliary muscle pulls the CB towards the lens. This slackens the zonule so that the elastic lens capsule is able to mould the lens into the accommodated shape.

c **False.** Numerous parasympathetic fibres supply the ciliary muscle.

d **False.** The posterior surface of the CB is smooth (pars plana). Its anterior surface is grooved.

e **False.** The lens grows by division of peripheral non-sheddable cells. The centre is its oldest part. The less is avascular and its cells apart from antena epithelium lose their nuclei and so cannot divide.

15 a **False.** Central sclerosis increases the refractive index of the lens and so can cause 'second sight' in the elderly. Some people can read more easily at 80 than at 65.

   b **False.** In senile cataract either the normal tendency of the central nuclear fibres is intensified ('nuclear cataract') or hydration and coagulation of proteins occurs mainly in the cortex ('cortical cataract').

   c **True.** Cataractous changes can be due to excessive local hydration, to calcium deposition, to protein coagulation, to trauma, to metabolic or viral conditions, or to local or systemic steroids.

   d **True.** Patches of atrophy in the iris can be due to a raised IOP and hence can suggest glaucoma. Note that a muddy iris, small irregular pupil, and sluggish reaction to light suggest iritis.

   e **False.** Iridonesis should always be investigated. It may be due to lack of support from a shrunken, subluxed, or absent lens.

16 a **False.** All the layers of the retina come from primitive forebrain, except the pigment layer.

   b **True.** Rods and cones exert a tonic inhibitory effect on bipolar cells in the dark. This is removed by exposure to light.

   c **True.** However, the overall cell ratio is bipolar:cones (1:5) and bipolar:rods (1:125).

   d **False.** Ganglion cells have slow spontaneous discharge even in the dark. The 'on' response means increased discharge in the light; the 'off' response means an increased discharge in darkness.

   e **False.** Clumps of 'on' cells are surrounded by 'off' cells and vice versa.

17 a **False.** Albinos also have abnormal crossing of fibres in the optic chiasma and an immature fovea.

   b **True.** Excess melanin can cause blindness, as in some cases of retinitis pigmentosa.

   c **False.** Typically the abnormal lipid in familial lipid degeneration disease is deposited in the ganglion cells.

   d **False.** The foveal and macular regions suffer most.

   e **True.** The potential space between these represents the primary optic vesicle.

18 a **False.** The postnatal vitreous has no blood vessels and is incapable of inflammation. It subserves only optical functions.

   b **False.** The posterior chamber is bounded by the iris in front and the lens, zonule, and CB behind.

   c **False.** A small amount of aqueous is reabsorbed in the posterior chamber as well as by the choroid and retina.

   d **True.** Also about 2–3 ml of aqueous is formed per minute.

   e **False.** Only about 5% of aqueous humour is formed by passive diffusion; 95% is actively secreted by the ciliary bodies.

19 a **True.** There is danger if the IOP exceeds 26 mmHg, or exceeds 22–25 mmHg when there is also optic disc or perimeter changes.

b **False.** Perimetry and tonometry are the best methods of detecting glaucoma. Note that photography is a more sensitive screening agent than perimetry, tonometry, or direct ophthalmoscopy in detecting diabetic retinopathy.

c **False.** Ophthalmopathy in Grave's disease is mainly due to the over-production of retro-orbital glycoaminoglycans, presumably from inter-action between them and anti-TSH receptor antibodies.

d **True.** The IOP is increased (spuriously) in the upgaze position because the inferior rectus muscle is often the most affected in Grave's disease.

e **True.** The ciliary muscle pulls the scleral spur, opens the local trabecular meshwork and so aids in the reabsorption of aqueous.

20 a **False.** Primary closed-angle glaucoma is more common in hypermetropes. The eye is smaller than normal, the lens more anterior, and the forward-tilted iris blocks the anterior angle.

b **True.** Dilated pupils in dim light allow light to enter via the clear periphery of lens when the centre may be opaque due to senile cataract.

c **False.** Closed-angle glaucoma occurs most often in middle-aged people. Note there is precipitation of angle closure by dilated pupils in the elderly watching TV in a dark room.

d **False.** A semi-dilated pupil offers the most resistance, especially in the hypermetropic eye.

e **True.** Excessive pressure builds up in the posterior chamber behind a semidilated iris after mydriatic drops.

21 a **True.** In the liver carotinoids are are converted to vitamin A which is dehydrogenated to retinene in the retina.

b **True.** The periphery has mainly rods while the fovea has only cones.

c **False.** Rods are far more sensitive to light than cones. Hence you have peripheral but not central vision in twilight.

d **False.** Rods react most at the green-blue end (500 nm); cones react most at the yellow-red end (550–600 nm).

e **True.** Night blindness is one of the very earliest symptoms of vitamin A deficiency. It also occurs early in retinitis pigmentosa. Mainly the periphery of the retina is affected.

22 a **False.** Cone pigments regenerate rapidly; hence there is only a very transient initial 'blindness' or dazzle on passing from dim to bright light.

b **False.** Dark adaptation takes about 30 min to become fully effective.

c **True.** Retinal alcohol dehydrogenase (a Zn-containing enzyme) is the same as that found in the gastric mucosa.

d **False.** Red goggles allow red light into the eye. Rods do not respond to red light. Thus red goggles protect dark adaptation and hence the use of red lighting in submarines.

e **False.** About 8% of males and 0.5% of females are colour blind. The condition is transmitted as an X-linked recessive state.

23 a **False.** The nodal point is in the posterior part of the true lens.

b **False.** After removal of the lens accommodation is lost and the eye is extremely hypermetropic (aphakia).

c **False.** After removal of the lens vision is usually uniocular due to suppression of the image from the more badly affected eye.

d **True.** The retinal images are only 2 μm apart (diameter of a cone).

e **False.** Snellen's types assume an average visual angle is 1 minute. Snellen's types are arranged so that each whole character occupies 5 minutes of arc at the stated distance.

24 a **True.** The lateral geniculate body has six well-defined laminae, three from each eye.

b **True.** Note: foveal maldevelopment in albinos causes great difficulty in accurate visual resolution.

c **False.** All parts of the retina are topographically aligned in the primary visual area. Macular representation is inordinately large, approximately ten times as great.

d **False.** The mid brain contains relay nuclei only and vision is not finally perceived there.

e **False.** Ipsilateral eye movements originate in the contralateral middle frontal gyrus.

25 a **False.** Disconjugate movement is normal in focusing on near objects.

b **False.** Saccadic movement describes all other voluntary conjugate movements. Smooth-pursuit is the term used to describe the following of objects (suppresses saccadic movement).

c **False.** Saccadic movements are generated in the pons (near REM neurons) and are executed by third, fourth, and sixth cranial nerves.

d **False.** Visual acuity is suppressed during saccadic movements (< 100 ms) but is normal during intervening steady periods.

e **True.** Damage to the medial longitudinal bundle causes dissociation of conjugate eye movements. Nystagmus and specific palsies from individual nuclear damage are also common.

26 a **True.** The consensual reflex depends on the decussation of fibres in the optic chiasm and on the integrity of the pretectal nuclei, i.e. an intact II nerve is needed.

b **False.** The unpaired Perlia (and pontine centres) mediate accommodation and convergence. The Edinger–Westphal nuclei control pupillary constriction.

c **True.** However, most commonly the afferent limb of the blink reflex is the trigeminal nerve (from the cornea). The optic nerve is the sole afferent limb of the dazzle response.

d **False.** Unconscious blinking is abolished by topical anaesthetics and by post-herpetic scarring of the cornea.

e **True.** This is because the effector muscles of blepharospasm are the orbicularis oculi and are innervated by the facial nerve.

27 a **True.** If one pupil dilates relative to the other then the optic nerve is damaged on the dilating side. Consider the Marcus Gunn or swinging-flashlight test.

b **True.** This is the classical bilateral Argyll Robertson pupil.

c **True.** A lesion of the ciliary ganglion causes unilateral loss of the light reflex but there is retention of the near reflex.

d **False.** Accommodation is probably evoked by a blurring of the perceived image (perception blurring).

e **False.** Pupil dilation in such cases is also due to inhibition of pupillary constrictor tone (Wernicke's reaction).

28 a **True.** Also sometimes large oscillations (hippus) are found in disease, e.g. in multiple sclerosis.

b **False.** Slow dilation of the pupil despite a constant light source may indicate optic neuritis. Sluggish oscillations are also typical.

c **False.** The parasympathetic, placed superficially in the oculomotor nerve, is the first affected by an irritative lesion.

d **False.** A progressive extradural haemorrhage usually causes initial dilation of the ipsilateral pupil because of parasympathetic damage (28c).

e **True.** A Horner's syndrome is caused by sympathetic paralysis to eye, upper limb, and face. This can be caused by a Pancoast tumour that erodes the stellate ganglion on the neck of the first rib, or by the tracking back of xylocaine along this rib when performing brachial plexus anaesthesia by Patrick's method.

13

Muscle

Contraction (Qs 1–5)

1 In skeletal muscle:

a Tropomyosin (Tm) is the part of myosin that binds to actin.
b Tropomyosin has a high affinity for calcium ions.
c Troponin (Tn) facilitates ATPase activity during muscle contraction.
d Troponin is crucial for keeping actin and myosin apart.
e The shape of the Tn molecule is crucial for contraction/relaxation.

2 In skeletal muscle contraction:

a Excitation–contraction coupling refers specifically to the role of calcium in the coupling of actin and myosin.
b An action potential in skeletal muscle lasts 1–2 ms.
c A maximal number of binding sites are made available in a single twitch.
d Maximal tension is developed in a single twitch (all-or-none rule).
e The lumen of the T-tubules is continuous with the ECF.

3 Within skeletal muscle:

a The sarcoplasmic reticulum (SR) becomes one with the T-tubules at the lateral sacs (terminal cisternae).
b Fast muscle fibres contain myosin with high ATPase activity.
c Red and pale fibres are named because of their mitochondrial content.
d Depleted muscle ATP levels correlate well with fatigue.
e Exercise facilitates the entry of glucose into muscle cells.

4 In smooth muscle:

a there is actin, myosin, tropomyosin, and troponin.
b Actin and myosin are arranged in a rosette fashion.
c the upstroke of action potential is due to Na^+ entry.
d T-tubules are well developed in smooth muscle.
e Receptors for neurotransmitters are located at single but poorly developed end plates.

5 In muscle metabolism:

a Glucose is the immediate source of chemical energy when muscle contracts.
b Lipid is the predominant fuel for muscles at rest.
c Muscle glycogen lasts about 10 h in moderate exercise.
d Anaerobic power only lasts for 4–5 min in very severe exercise.
e In normal people there is no pulmonary limitation to aerobic performance.

Contraction (Answers)

1 a **False.** Tropomyosin keeps actin and myosin apart.
   b **False.** Troponin C has a high affinity for calcium ions.

c **True.** At rest troponin I inhibits ATPase. This inhibition is removed during contraction.

d **True.** Troponin T facilitates tropomyosin in this action.

e **True.** Ca binding produces a change in the shape of troponin which affects Tm and exposes myosin-binding sites on actin. The removal of calcium does the reverse (relaxation).

2 a **False.** Excitation–contraction coupling refers to the events by which an AP leads to actin/myosin cross-bridge formation.

   b **False.** An AP in skeletal muscle lasts 2–4 ms, travels at about 5 m/s, and is completed before mechanical events begin.

   c **True.** A single AP releases enough calcium to saturate Tn so that all myosin-binding sites on actin are initially available.

   d **False.** Binding sites are not available long enough for maximal tension to develop in a single twitch. Contrast this to tetanus where successive APs flood the myoplasm with calcium ions.

   e **True.** Sarcolemma AP is conducted deep into the muscle fibre via T-tubules.

3 a **False.** T-tubules cross between adjacent lateral sacs. AP in T-tubules releases calcium, which diffuses into cytoplasm, from sacs.

   b **True.** The contraction to peak tension time is only 40 ms in a fast fibre but 100 ms in a slow one (lower ATPase activity).

   c **False.** Myoglobin gives muscle its red colour. Slow endurance fibres contain a high myoglobin and many mitochondria.

   d **False.** The cause of muscle fatigue is uncertain but may be related to one or more of the following: H^+ ion, lactate, PO_4 rise, failure of the Ca^{2+}-ATPase of SR.

   e **True.** Exercise facilitates glucose entry into muscle cells possibly by increasing the number of glucose transporters (specifically GLUT-4) translocated from myoplasm to T-tubule and sarcolemma.

4 a **False.** Smooth muscle does not contain troponin. Ca^{2+} binds with calmodulin.

   b **True.** Myosin is interspersed among actin filaments in smooth muscle.

   c **False.** Upstroke of AP in smooth muscle is due to entry of Ca^{2+}.

   d **False.** Smooth muscle has no T-tubules; the SR is absent or poorly developed.

   e **False.** Receptors are located throughout the smooth muscle membrane. No specific end plates are found.

5 a **False.** The ATP–ADP–phosphocreatine (PCr) system provides immediate energy for contraction in all types of muscle.

   b **True.** Lipid is also the main fuel source in light exercise.

   c **False.** In moderate exercise in reasonably fit people muscle glycogen only lasts 3–4 h.

d **False.** In severe exercise anaerobic energy contributes 100% (> 4000 W) in the first few seconds, 10% at 5 min, and 5–10% up to 60 min (exhaustion).

e **True.** At maximal O2 consumption (VO_2max) it is still possible to increase breathing voluntarily.

Exercise (Qs 6–11)

6 Respiratory changes in exercise include:

a blood gases not altering during light to moderate exercise.

b hyperventilation, which is mediated by the carotid bodies in severe exercise.

c athletes achieving a minute ventilatory volume (MVV) > 200 litres.

d increased O_2 consumption (VO_2) to about 3.5 l/min in untrained people.

e the respiratory quotient (RQ) reaching 2.0 in severe exercise.

7 Cardiovascular adjustments in exercise are:

a In mild to moderate exercise the CVP falls.

b In severe exercise the end diastolic volume (EDV) increases.

c In moderate exercise the coronary blood flow (CBF) remains the same % of cardiac output as at rest.

d In mild to moderate exercise the skin blood flow remains the same % of cardiac output as at rest.

e Total peripheral resistance (TPR) remains about normal in moderate exercise.

8 Changes brought on by exercise:

a Core temperature will not rise by more than 1 °C in moderate exercise.

b Plasma levels of cortisol rise in exercise.

c Opioids are suppressed during moderate exercise.

d ECF formation is dramatically increased in exercise.

e Recumbency is the most physiologically advantageous posture to adopt after moderate to severe exercise.

9 Training can result in:

a muscle strength doubling in intensive training.

b individual muscle fibres splitting down the middle in the process of 'getting fit'.

c Hb concentration rising as fitness increases.

d Gonadotrophin secretion rising in male athletes.

e Histological fibre type changing as appropriate for the particular exercise (isometric, isotonic, or endurance).

10 *The athlete's heart:*

a   The resting cardiac output of an athlete and of a matched untrained person is about the same.

b   Cardiac hypertrophy is similar in an endurance runner and in a sprinter (the hare and the rabbit).

c   The greater left ventricular (LV) volume is due mainly to a larger LV cavity.

d   LV wall thickness should not exceed 13 mm.

e   Cardiomegaly can be a response to, rather than a cause of, bradycardia.

11 *Benefits of exercise include:*

a   physical fitness, which is a long-term protector against CVS disease.

b   benefits to patients with obstructive airway disease where armwork is more beneficial than legwork.

c   benefits accrued more from the intensity than the quantity of exercise.

d   beneficial alteration of the HDL:LDL ratio by lowering LDL levels.

e   in those with coronary heart disease, the functional work capacity of the heart, if the exercise is regular.

Exercise (Answers)

6   a   **False.** In light to moderate exercise the arterial gases remain unaltered; the venous ones change predictably.

b   **True.** Carotid bodies are strongly activated by the gross alterations in PaO_2 and pH of arterial blood.

c   **True.** An MVV of 245 l/min was achieved by a UK heavyweight rower.

d   **True.** The VO_2 max. of elite athletes averages 5.26 l/min.

e   **True.** RQ rises markedly in severe exercise due to the excess CO_2 produced by interaction of lactic acid and sodium bicarbonate.

7   a   **True.** CVP falls due to the increased stroke volume and heart rate.

b   **True.** The EDV increases from about 145 ml to over 200 ml. This augments the cardiac output from what is literally a failing heart.

c   **True.** Resting and exercise CBF are both about 4% of cardiac output.

d   **False.** Skin blood flow rises from 9% (500 ml) at rest to 11% (19 000 ml) in exercise.

e   **False.** TPR falls about 50% in moderate exercise. The increased cardiac output raises the systolic BP and also helps maintain diastolic pressure at near normal.

8   a   **False.** In exercise the core temperature may rise to over 40 °C without discomfort.

b   **True.** Exercise is stressful, thus all stress hormones rise.

c   **False.** Opioid release is greatly increased during exercise, shock, surgery, and pain, hence the addiction of training.

d **True.** Over 1 litre of ECF may be 'lost' into active muscles because of raised capillary pressure and increased capillary permeability. Decreased ECF formation elsewhere helps avoid harmful haemo-concentration.

e **False.** Recumbency reduces FRC (functional residual capacity) and pools blood in lungs. A semistooped stance is best, with hands placed firmly above knees to stabilize upper chest and facilitate inspiration by allowing use of accessory muscles of respiration.

9 a **False.** Muscle strength increases up to 30% in a 6–8 week programme. Little improvement occurs after that.

b **True.** In training muscle fibres may split in half and increase in diameter (30–60%).

c **False.** Hb concentration falls, possibly from damage to RBCs in passage through exercising areas. This is known as 'haemolytic anaemia' of exercise.

d **False.** Gonadotrophin secretions fall in both males and females.

e **False.** The metabolic profile of muscle fibres changes appropriately but the histological profile remains unaltered.

10 a **True.** This is because the athlete compensates for a larger stroke volume by having a slower heart rate.

b **False.** The endurance athlete has larger heart and slower heart rate than the untrained individual.

c **True.** Both the ventricular wall and cavity increase in size in proportion to each other (concentric hypertrophy).

d **True.** Hypertrophic cardiomyopathy is likely when the LV wall is thicker than 13 mm.

e **True.** The statement is true. Many hold that bradycardia (increased vagal tone) is a response in the athlete to cardiomegaly and consequent increase in cardiac output.

11 a **False.** Physical fitness is a long-term predictor of, but an unproved long-term protector against, CVS disease

b **False.** Armwork is damaging. Muscles of shoulder girdle (involved in armwork) are also involved in breathing in these patients. Armwork is damaging and poorly tolerated as it involves and desynchronizes muscles of the shoulder girdle needed for ventilation in these patients. Legwork is better tolerated and provides a large muscle 'sump' to handle H^+ and lactate, which might otherwise overtax the respiratory capabilities.

c **True.** It is generally accepted that exercise should be sufficient to produce sweating or tachypnoea or a heart rate 60–80% of maximum for a period of greater than 20 min.

d **False.** Exercise chiefly increases the HDL cholesterol in the plasma.

e **True.** Regular exercise lowers heart rate and systolic BP so that the myocardial oxygen consumption is reduced.

Death in sport (Q. 12)

12 Sudden death in sport (SDS):

a Mitral valve prolapse is a common cause of SDS.
b Rupture of the right ventricular wall is a well-documented cause.
c Coronary artery disease is the most common cause if over 40 years.
d Ruptured aortic aneurysm is the most common cause if the athlete is aged < 40 years.
e Hypertrophic cardiomyopathy is associated with a very large LV cavity.

Death in sport (Answer)

12 a **False.** However, it is the most common valvular abnormality in the UK and USA.

b **True.** This occurs in Uli's syndrome (parchment heart and paper-thin RV wall).

c **True.** Coronary artery disease, including anomalous origin of the coronary arteries and congenital abnormalities of the coronary artery tree, are well-known causes of sudden death.

d **False.** Hypertrophic cardiomyopathy is the most common, followed by RV dysplasia, Marfan's syndrome, aortic stenosis, and brain haemorrhage. Commotio cordis ('concussion of the heart') with electrical disturbance and without any organic heart disease is probably important in younger subjects after direct blunt injury to the chest.

e **False.** In hypertrophic cardiomyopathy the LV cavity is reduced. Hypertrophy of the wall reduces compliance and so impairs diastolic filling.

Muscle disease (Q. 13)

13 In muscle disease:

a Creatine kinase is low in Duchenne-type muscular dystrophy.
b Weakness and degeneration of muscle fibres in dystrophy is often associated with an abnormal sarcolemmal protein.
c Myotonic muscle disorders have increased resting muscle tone.
d Abnormalities of the triad system (two lateral sacs, one T-tubule) are key elements in many cases of myotonia.
e Blockade of the neuromuscular junction with curare abolishes myotonia.

Muscle disease (Answer)

13 a **False.** Creatine kinase is usually raised, often to quite high levels, in the Duchenne type of dystrophy.

b **True.** Dystrophin, the product of the gene affected in many dystrophies, is located in the sarcolemma. Its function is to anchor actin to the cell membrane.

c **False.** A characteristic of myotonic muscle disorders is a delayed relaxation after voluntary contraction or after mechanical stimulation.

d **False.** Myotonia is a disorder of muscle membrane excitability (increased Na^+ and/or decreased Cl^- conductance, protein kinase defect with altered phosphorylation of channel proteins).

e **False.** Curare does not affect ion conductance in the plasma membrane.

14

The final test 1

Chapters 14 and 15 are two exams, each comprising 20 questions. Several of the questions in these final tests are more difficult than you are likely to meet in your exam, so you will find this exercise a real challenge. Score yourself out of 100. Score 1 point for each correct answer, deduct 1 for each incorrect answer, give zero for a no attempt. A pass is 60 points. Good luck!

1 Glutamine ($(COOH)_2.(CH_2)_2.CH–NH_2$):

a The lung is a primary site for net glutamine synthesis.

b The kidney is normally the major site of glutamine utilization.

c In cells where the usage of glutamine is high, most of the glutamine is used as a respiratory substrate.

d There is a large net absorption of glutamine from the intestine (normal diet).

e Liver glutaminase is different from that found in other tissues.

2 Assume ICF is 25 litres; ECF, 15 litres; osmolality, 300 mosmol/kg:

a Once water is added to the bloodstream it takes about 30 min for its full distribution between ECF and ICF.

b If the ECF is expanded by 10 litres of water, then the ICF osmolality falls to 260 mosmol/kg (after osmotic equilibrium).

c If isotonic saline is added to ECF, the only immediate effect is an increase in ECF volume (after osmotic equilibrium).

d Infuse 2 litres of 4.4% NaCl: both ECF and ICF will expand to reach the same volume (after osmotic equilibrium).

e Hypertonic glucose, sucrose, or mannitol infusions only reduce ICF volume for a few hours.

3 In the kidneys:

a medullary rays are groups of collecting ducts converging on the renal papillae.

b medullary rays are supplied by venous blood ascending from the renal pelvis.

c the PO_2 in the renal cortex is about 50 mmHg.

d the PO_2 of the medulla is in the range 30–40 mmHg.

e loop diuretics decrease the PO_2 in the medulla.

4 Plasma albumin:

a is virtually the only plasma protein not a glycoprotein.

b attaches readily to sialic acid.

c has anticoagulant properties.

d falls in sepsis, trauma, and major surgery due to an increased catabolism and/or reduced albumin synthesis.

e absolute level in the plasma is critical for the formation of oedema in cases of hypoalbuminaemia.

5 The volume of cells:

a can only be regulated by the loss or gain of osmotically active solutes.

b is reduced chiefly by the extrusion of NaCl followed by water.

c is increased chiefly by the uptake of NaCl and KCl.

d is significantly determined by the cytosolic concentration of polyols, e.g. sorbitol and myo-inositol (organic osmolytes).

e is reduced in the brain by infusions of hypertonic mannitol along predictable osmotic lines in cases of cerebral oedema.

6 Pathophysiology of the liver:

a First-pass metabolism is increased in the elderly (slow transit time).

b The liver is an important reservoir of blood in hypovolaemic shock.

c Ascites is common once portal hypertension occurs.

d Hypoalbuminaemia is an important initiating aetiological factor in cirrhotic ascites.

e Underfilling of the circulation and consequent renal retention of NaCl and H_2O are critical for the development of ascites in cirrhosis.

7 In liver disease:

a α1-Antitrypsin normally comprises most of the α1 peak on electrophoresis.

b Cerebral oedema in hepatic encephalopathy is usually caused by hypervolaemia and/or hypoalbuminaemia.

c Cerebral oedema patients often have hypotension and bradycardia.

d In acute liver failure elevated levels of benzodiazepines may be found in the brain.

e Acidosis is common in acute liver failure.

8 Due to lung mechanics:

a Obstructive airway disease is associated with a positive mean intrapleural pressure.

b A large functional residual capacity (FRC) tends to reduce the hydrostatic gradient for the accumulation of tissue fluid in lung.

c Small to moderate degrees of positive end expiratory pressure (PEEP) are created in chronic obstructive airway disease (COAD), even at rest.

d A high lung volume reduces pulmonary vascular resistance.

e Pulse amplitude during the strain phase of the Valsalva manoeuvre is related to the pulmonary capillary wedge pressure (PCWP).

9 Normobaric oxygen (100% O_2) therapy:

a is of great value to subjects with atmospheric hypoxia (hypoxic hypoxia).

b can correct the deranged blood gases in hypoventilation hypoxia.

c is of little or no use in hypoxia from a diffusion block, e.g. pulmonary oedema.

d is of value in anaemic hypoxia.

e is of great value in hypoxia from abnormal intrapulmonary shunting.

10 Thermoregulation:

a The concentration of urea in sweat at low sweat flow rates is roughly the same as in plasma.

b Sweat is not produced if the sweat duct is occluded by a pressure greater than the systolic blood pressure.

c Excitation of α 1-adrenoreceptors stimulates sweating.

d Radiation is still a useful factor in getting rid of excess heat even when the body temperature exceeds ambient temperature.

e The cooling effect of wind at low velocities is directly proportional to the wind velocity (for a standard set of circumstances).

11 Electrical activity in heart:

a The P wave on the ECG is due to inward Na^+ currents, and the T wave is largely due to closure of Na^+ channels.

b In slow-response fibres (SA and AV nodes) depolarization during phase 0 is primarily dependent on Ca^{2+} entry into the cells.

c A rising cytosolic Ca^{2+} inhibits outward passage of K^+ in ordinary cardiac muscle cells.

d Outward-rectifying K^+ currents are promoted by cytosolic Na^+.

e Inward-rectifying K^+ current channels close with depolarization.

12 In cardiovascular physiology:

a Sympathetic stimulation accelerates the rate of Ca^{2+} reuptake by the sarcoplasmic reticulum (SR) in vascular myocytes.

b A sudden haemodynamic overload increases the cardiac output.

c There is a clear relationship between death from pulmonary embolus and the rate of deep venous thrombosis in elective orthopaedic surgery.

d Myocardial magnesium concentrations are relatively low during acute cardiac ischaemia.

e Magnesium vasodilates both coronary and peripheral vessels.

13 Control of skeletal muscle circulation:

a Distension of the resistance blood vessels by intravascular pressure contibutes significantly to the basal vascular resistance in non-active muscle.

b Sympathetic nerves are in contact with all layers of vascular smooth muscle within the muscle.

c The arteriolar resistance vessels contain β 2 subtype-adrenoreceptors which mediate vasodilation.

d   There is a limited cholinergic vasodilator outflow from the lumbar region to lower limb muscles.

e   Constriction of muscle resistance vessels depends solely on activity in noradrenergic nerves.

14 *Regarding nitric oxide (NO):*

a   Its actions are mimicked by monomethyl-L-arginine.

b   It is very unstable when dissolved in water.

c   Normally there is a small amount of NO in exhaled air.

d   In the body most is ultimately oxidized to nitrogen dioxide.

e   Nitrogen dioxide is harmless to normal cells.

15 *Physiological functions of NO:*

a   NO is essential for normal vasodilator tone.

b   Arteries are a better source of NO than veins.

c   NO formed by platelets causes feedback antiaggregation.

d   NO inhibits glutamine in the CNS.

e   Nitregeric nerves usually contract smooth muscle sphincters.

16 *In normal endocrine function:*

a   Peptide hormones are seldom stored within their cell of origin.

b   Iodide is the ionic form of iodine in the body.

c   Cholesterol is the precursor of all steroid hormones.

d   In men adrenal androgens are as important as testicular testosterone.

e   'Short-loop' feedback refers to inhibition of a pituitary tropic hormone by its target hormone, e.g cortisol inhibits ACTH.

17 *Thyroid function:*

a   Circulating T_4 (as well as T_3) inhibits TSH and TRH production.

b   Thyroid hormone stimulates (upregulates) TRH receptors.

c   In hypothyroidism from thyroid disease TSH is usually elevated.

d   An elevation in TSH is usually found in hyperthyroidism.

e   TRH causes a marked rise in TSH in hyperthyroid states.

18 *Concerning insulin:*

a   It stimulates the transport of K^+ ions.

b   It inhibits ovarian androgen production.

c   Obesity is associated with a defect in the kinase activity of the insulin receptor.

d   Resistance to its actions may be caused by mutation of the insulin molecule, e.g. a failure to cleave 'abnormal' proinsulin

e   Patients with resistance to the glucose-lowering effects of insulin commonly exhibit signs of other insulin effects.

19 *Within the CNS:*

a Synapses in the cerebral cortex are the most vulnerable to general anaesthetics.

b The spinal distribution of μ opioid receptors parallels that of encephalins.

c The action potential in a nerve originates on the plasma membrane of the soma.

d Nerve cell lysosomes are membrane-bound vesicles.

e CSF has roughly the same concentration of Cl^- as plasma.

20 *The vestibulo-ocular reflex (VOR):*

a is initiated by photorceptors in the retina.

b is a monosynaptic postural reflex.

c The visual system can stabilize retinal images even if the VOR is absent.

d The first-order neurons of the VOR go to the oculomotor nuclei.

e Eye motor events and retinal images influence the VOR.

Answers

1 a **True.** Lung, brain, adipose tissue, and in some conditions the liver are primary sites for net glutamine synthesis.

b **False.** The small intestine, macrophages, and lymphocytes are major users of glutamine. The kidney is only a major user in acidosis.

c **True.** Glutamine is the most common amino acid in the body. Over 5% is used for biosynthesis, e.g. purines and pyrimidines.

d **False.** Most body glutamine is synthesized *de novo*. Note that the GIT uses very large amounts of glutamine.

e **True.** Liver-type glutamine feeds NH_3 directly to the urea cycle and is activated by NH_3.

2 a **False.** Osmosis begins immediately at cell membranes. Full distribution among all fluid spaces is complete within minutes.

b **False.** Osmolality of both compartments falls to 240 mosmol/kg.

c **True.** Note that infusions of hyper- or hypotonic saline will cause fluid shifts.

d **True.** Both increase to 21 litres. Both achieve osmolality of 357 mosmol/kg.

e **True.** Osmotic dehydration of ICF lasts for about 2–4 h. Glucose is metabolized, and sucrose and mannitol are excreted rapidly by the kidney. Osmotic diuresis ensues.

3 a **False.** Medullary rays are finger like projections of the medulla into the cortex. They are devoid of glomeruli.

b **True.** Thus medullary rays have a very low oxygen supply.

c **True.** There is a very high O_2 usage in the renal cortex.

d **False.** The PO_2 of the renal medulla is between 10 and 20 mmHg (because of A-V transfer of O_2 in vasa recta).

e **False.** Loop diuretics increase the PO_2 by inhibiting active transport in the thick ascending segment of Henle's loop.

4 a **True.** The vast majority of plasma proteins are glycoprotein and contain CHO attached at an asparagine residue.

 b **False.** Sialic acid forms the terminus of CHO residues of plasma glyco-proteins.

 c **True.** Plasma albumin inhibits platelet aggregation and enhances the anticoagulant effects of antithrombin III.

 d **False.** The fall in plasma albumin occurs too quickly for this to be the case. A redistribution of albumin is more likely.

 e **False.** The transcapillary gradient of albumin, not its absolute level in plasma, is important in the genesis of oedema.

5 a **True.** Cell volume is regulated primarily by the passage of inorganic ions in and out of the cell.

 b **False.** Regulatory decrease in cell volume is chiefly due to the opening of separate K^+ and Cl^- channels or to activation of the K^+/Cl^- transporter.

 c **True.** NaCl and KCl accumulate in the cell because of activation of Na^+/H^+ and Cl^-/HCO_3^- exchangers or the $Na^+/K^+/2Cl^-$ cotransporter.

 d **True.** Polyols, amino acids, and methylamines are present in tens to hundreds of mmol/l in the cytoplasm of all cells.

 e **False.** Hypertonic mannitol may reduce intracranial volume by reducing the haematocrit, causing cerebral vasoconstriction, and/or inhibition of the cellular uptake of Na^+, K^+, and Cl^- which is the normal response of cells exposed to a hypertonic environment.

6 a **False.** First-pass metabolism is decreased in the elderly (liver inefficient) and in liver disease. Lower effective drug doses are needed in such cases.

 b **True.** The liver can add about 350 ml blood in acute hypovolaemia.

 c **False.** Ascites only occurs if there is severe portal hypertension.

 d **False.** Hypoalbuminaemia compounds cirrhotic ascites but is not causative.

 e **False.** Both under- and overfilling theories are in doubt. Current thinking implicates sympathoadrenal activity, peripheral arteriolar dilatation, and multiple A-V fistulae as primary events.

7 a **True.** α 1-Antitrypsin comprises 90% of the α1 peak.

 b **False.** Cerebral oedema is caused by a loss of cell membrane integrity and increased permeability of the blood–brain barrier.

 c **False.** Usually patients with cerebral oedema have hypertension and brady-cardia (Cushing's reflex).

 d **True.** Endogenous benzodiazepines enhance the impending stupor/coma of acute liver failure.

 e **False.** A CNS-induced respiratory alkalosis usually occurs in liver failure.

8   a   **False.** In obstructive lung disease there is a more negative mean intrapleural pressure. This is due to persistently greater inspiratory efforts. There is a relatively small increase in expiratory intrapleural pressure.

    b   **False.** A large FRC leads to a more negative intrapleural pressure which in turn causes an increased transmural pressure in pulmonary vessels. This increases the hydrostatic gradient and favours oedema formation.

    c   **True.** Auto-PEEP is more marked in coughing and in exercise. Expiratory narrowing of the glottis may be important in causing this.

    d   **False.** A high lung volume increases pulmonary vascular resistance but lowers airway resistance.

    e   **True.** Pulse amplitude variations in the Valsalva manoeuvre provide a non-invasive way to monitor the PCWP in heart failure.

9   a   **True.** O_2 will elevate the depressed PaO_2 to normal levels in patients with hypoxic hypoxia.

    b   **False.** O_2 therapy provides no benefit for the associated hypercapnia. Note that there are disappointing results in spite of maintaining a supranormal cardiac index and a normal SvO_2 in critically ill patients.

    c   **False.** An increased O_2 alveolar–capillary gradient drives considerably more O_2 into pulmonary capillary blood.

    d   **True.** 100% inspired O_2 increases the alveolar PO_2 (PAO_2) to about 600 mmHg and increases the PaO_2 so that 1.5 ml O_2 is now dissolved in 100 ml of plasma.

    e   **False.** Blood passing via 'good' alveoli is already fully saturated with oxygen (although it can carry some extra oxygen in solution). Increasing O_2 does not change the size of the shunt.

10 a   **False.** Urea is concentrated in sweat (especially at low flow rates) because of disproportionate water reabsorption (sweat is hypotonic).

    b   **False.** Sweat secretion can produce an intraductal pressure of up to 250 mmHg.

    c   **True.** Furthermore α 1-receptors are present in erector pilae muscles.

    d   **False.** Body gains heat by radiation when the body–environment temperature gradient is reversed.

    e   **False.** The cooling effect of wind is proportional to the square root of low to moderate velocity wind flow.

11 a   **False.** Rapid propagation of the wave of depolarization generated by opening of Na^+ channels causes both P wave and QRS complex, but the T wave (repolarization) is due largely to the outward flow of K^+.

    b   **True.** Nodal upstroke with little or no overshoot is carried by the entry of Ca^{2+} via long lasting (L-type) Ca^{2+} channels that are dihydropyridine sensitive. Contrast this with the predominantly Na^+-dependent depolarization of fast response atrial and ventricular fibres.

    c   **False.** Cytosolic Ca^{2+} promotes repolarization by activating an outward-rectifying potassium current.

 d **True.** ATP, ACh, acidosis, and arachidonic products also affect this.

 e **True.** Closing of the inward-rectifying K^+ current channels decreases repolarizing currents and prolongs the AP plateau. In resting myocardial cells these channels maintain the transmembrane potential.

12 a **True.** Thereby sympathetic activity promotes relaxation (lusitropy).

 b **False.** A sudden fluid overload may cause acute heart failure with a low cardiac output, left ventricular dilatation, and pulmonary congestion.

 c **False.** The overall death rate in unprotected (no anticoagulants) orthopaedic patients is much lower than was formerly thought.

 d **True.** Mg^{2+} has been widely accepted as a cardioprotective agent.

 e **True.** Mg^{2+} is also an antiarrhythmic drug and antiplatelet agent.

13 a **True.** Intrinsic basal vascular resistance is largely due to the direct action of stretch, especially of pulsatile stretch on smooth muscle.

 b **False.** Sympathetic fibres are only in contact with the outer layers of vascular smooth muscle.

 c **True.** The β 2-adrenoreceptors in arterioles respond chiefly to circulating adrenaline from the adrenal medulla.

 d **True.** Cholinergic sympathetic vasodilators almost certainly exist in man and innervate arterioles, especially in the lower limbs. This probably causes NO release and inhibits NA release (like adenosine, histamine, K^+, serotonin, and hyperosmolarity).

 e **False.** Constriction of muscle resistance vessels is controlled solely by NA from nerves and by muscle compression.

14 a **False.** Monomethyl-L-arginine is a competitive inhibitor of nitric oxide and is a potent *in vitro* vasoconstrictor.

 b **False.** NO is stable in water. It is inactivated in seconds by the Fe^{2+} (ferrous iron) of haem.

 c **True.** NO is a gas. It is present in exhaled air in the range 5–20 p.p.b.

 d **False.** Most is rapidly oxidized to nitrites and then to plasma nitrate (30 mmol/l), which is finally excreted in the urine.

 e **False.** NO_2 is cytotoxic. NO forms NO_2 in air.

15 a **True.** NO made by microvascular endothelial cells is a principal determinant of resting vascular tone. NO acts via the cGMP system by combining with the Fe^{2+} of guanylate cyclase.

 b **True.** The somewhat greater concentration of NO in arteries may explain why arterial grafts remain patent more often than venous grafts.

 c **True.** Platelet NO reinforces the antiaggregation effects of endothelial NO and prostacycline.

 d **False.** Glutamine (Gln) generates more Gln during high-frequency synaptic transmission: Gln $\Rightarrow$ NMDA receptors $\Rightarrow$ Ca^{2+} $\Rightarrow$ NO synthases $\Rightarrow$ NO $\Rightarrow$ more Gln.

 e **False.** NO generally relaxes smooth muscle including sphincters, e.g. pylorus and lower oesophageal sphincter.

16 a **False.** Peptide hormones are usually stored in secretory vesicles. Exocytosis is dependent upon a rise in cytosolic Ca^{2+}.

  b **True.** Plasma iodide (I^-) is converted to iodine in the thyroid gland.

  c **True.** Cholesterol, whether delivered by plasma lipoproteins or made *in situ* is the major precursor of steroid hormones.

  d **False.** Adrenal androgens are of little importance in males. They may enhance sex drive in females.

  e **False.** 'Short-loop' feedback refers to inhibition of a hypothalamic releasing hormone by its target hormone, e.g. ACTH inhibits CRH.

17 a **True.** The pituitary and hypothalamus contain an unusually active 5-deiodinase allowing negative feedback from T_4.

  b **False.** T_3 downregulates TRH receptors.

  c **True.** TSH is usually low if the problem is in the pituitary or the hypothalamus.

  d **False.** Hyperthyroidism (elevated T_3 and T_4) is usually due to a thyroid-stimulating immunoglobulin or to gland autonomy. A low TSH then results.

  e **False.** In hyperthyroidism the elevated T_3 and T_4 will block TRH and TSH release. Conversely TRH will cause a TSH surge in hypothyroidism of thyroid gland origin.

18 a **True.** Insulin stimulates the passage of K^+ into cells in general, and in particular into muscle cells.

  b **False.** Insulin stimulates ovarian androgen production. Note: there is hirsutism in patients with hyperinsulinaemia.

  c **True.** A defect in the function of the insulin receptor may be the result of the abnormal metabolic milieu rather than the cause of insulin resistance.

  d **False.** Patients with a mutation of the insulin molecule respond normally to exogenous insulin.

  e **True.** There is hyperglycaemia because of resistance to the glucose-lowering effects of insulin (see acanthosis nigricans and ovarian thecal hyperplasia, Chapter 8 Q41a) but other actions of insulin are unimpaired.

19 a **False.** The most vulnerable synapses to GA are in the ventrolateral thalamus.

  b **True.** The μ-receptors are those most likely to be associated with descending pain pathways.

  c **False.** The AP originates in the axon hillock and initial 50–100 mm of the axon.

  d **True.** Lysosomes are membrane bound in all cells. The vesicles arise from the Golgi apparatus, contain hydrolytic enzymes, and are about 8 nm in diameter.

  e **False.** CSF contains 123–128 mmol/l of Cl^-; plasma has 95–105 mmol/l. CO_2 passes into glial cells where it is hydrated to H_2CO_3 (carbonic anhydrase present). The resulting HCO_3^- is exchanged for plasma Cl^-.

20 a **False.** The VOR is initiated in the vestibular apparatus. Any movement of the head stimulates this reflex.

b **False.** The shortest route for the VOR involves two synapses. There is only a 12 ms delay from head movement to eye motor response.

c **False.** Only the VOR can stabilize retinal images. The VOR creates reflex eye movements opposite in direction, equal in displacement, and equivalent in rate for every movement of the head.

d **False.** In the VOR the first-order neurons go to vestibular, cerebellar, and reticular nuclei; second-order go to oculomotor nuclei.

e **False.** The VOR is an open loop independent of visual feedback.

15

The final test 2

1 Selected biogenic compounds:

a The Tamm–Horsfall protein is secreted by the prostate.
b Actin is released from poorly perfused cells into the circulation.
c Fibronectins are normally only found in submucosal and subepithelial tissue.
d Many studies have shown a decrease of myoinositol in nerves affected by diabetic neuropathy.
e Somatostatin (and its analogue octreotide) reduce splanchnic/portal blood flow.

2 Passive movement of material in the body:

a Diffusion time increases in direct proportion to the distance over which a molecule diffuses.
b Lipid-soluble substances diffuse faster across the plasma membrane than they diffuse in solution in the ICF or ECF.
c Ions diffuse very slowly through the bilipid layer of the plasma membrane.
d 'Facilitated' diffusion is a saturable process.
e 'Simple' diffusion is a saturable process.

3 In the renal system:

a Medullary hypoxia is essential for efficient urinary concentration.
b In severe medullary hypoxia the thick segment of the tubule suffers most damage.
c Local PGE is a potent vasodilator in the renal medulla.
d Local PGE inhibits active transport in the thick ascending limb.
e Adenosine diverts blood from cortex to medulla.

4 Concerning plasma proteins:

a Factor Xa is the sole known physiological activator of prothrombin.
b Synthesis and release of acute-phase reactants (APRs) is solely by cytokines.
c Hepatocytes, B lymphocytes, and endothelial cells are the sole source of plasma proteins, apart from hormones in transit.
d Binding to albumin is the only physiological method for the transport of free fatty acids in plasma.
e There is more intravascular albumin than extravascular albumin.

5 Concerning the liver:

a The human liver weighs about 2.5 kg.
b The hepatic artery nourishes the hepatocytes.
c The hepatic artery can provide up to 50% of liver blood flow.
d Portal venous blood is sterile.
e The pressure in the portal vein is normally only 20 mmHg.

6 Further points about the liver:

a The hepatic clearance of Bromsulphalein (BSP) is useful in diagnosing the Dubin–Johnson syndrome.

b The volume of blood entering the liver exceeds the volume that leaves it.

c The A-V difference of indocyanine green can be used to determine the hepatic blood flow.

d Fatal hypoglycaemia follows a few hours after sudden occlusion of the circulation to the liver.

e The liver contributes approximately 75% of the albumin found in plasma.

7 About hypo- and hyperglycaemia:

a The rate at which the plasma glucose falls influences the occurrence of the symptoms and signs of hypoglycaemia.

b Adrenaline is necessary for the effective counterregulation of hypoglycaemia.

c A 5-hour glucose tolerance test (GTT) is an excellent test for excluding hypoglycaemia as a cause of symptoms.

d Hyperglycaemia causes a rise in cellular sorbitol.

e Advanced glycosylation end products (AGEs) are formed within tissue cells in hyperglycaemia.

8 About the respiratory system:

a Dead-space ventilation (V_d) is normally 20–30% of total minute ventilation (V_t).

b As V_d/V_t increases the expired PCO_2 ($PeCO_2$) rises above the $PaCO_2$.

c Normally at least 95% of the cardiac output should participate in gas exchange.

d The shunt fraction determines the ability of inspired O_2 to increase the PaO_2.

e The alveolar PO_2 (PAO_2) is proportional to inspired O_2 (PIO_2) and inversely proportional to the alveolar (arterial) $PaCO_2$.

9 Hyperbaric oxygen:

a increases oedema in postischaemic muscles.

b decreases red blood cell flexibility.

c is often of great use in air or gas embolism.

d increases white cell killing of aerobic bacteria and some fungi.

e stimulates new capillary and collagen formation in irradiated tissue.

10 Cardiac afferent nerves:

a Cardiac sympathetic afferents inhibit contractility of the heart.

b Isoproterenol increases firing of cardiac vagal afferents.

c Activation of cardiac vagal afferents increases in heart failure.

d Digitalis stimulates vagal afferents from heart.

e Vasopressin sensitizes cardiac vagal afferents.

11 *Plasma membrane ionic currents:*

a   Depolarization signals the h- and m-gates of Na^+ and Ca^{2+} channels to open.
b   In the closed, resting state both h- and m-gates are closed.
c   Closure of h-gates favours a closed inactive state.
d   Depolarization of the A-V node depends on fairly slow opening of small Ca^{2+} channels.
e   Depolarization of the His–Purkinje system is effected by opening of Na^+ channels.

12 *About hypertension:*

a   It is unusual for essential hypertension to start over 50 years of age.
b   The degree of BP elevation correlates closely with end-organ damage.
c   Essential hypertension is the most common cause of a hypertensive crisis (accelerated malignant hypertension).
d   Antihypertensive drugs reduce the risk of a hypertensive crisis.
e   Malignant hypertension is more common in higher socio-economic groups.

13 *Cholesterol:*

a   A raised plasma cholesterol is associated with an increased incidence of atherosclerosis.
b   Lowering a raised plasma cholesterol, especially a raised LDL fraction, reduces the incidence of coronary heart disease (CHD) in otherwise normal people.
c   Statin-type drugs cause an increase in non-cardiovascular deaths.
d   The plasma cholesterol rises as the level of dietary cholesterol rises.
e   Lowering the dietary cholesterol can lower the plasma cholesterol.

14 *Nitric oxide, pathophysiology:*

a   The lung and the liver are the major sources of inducible NO in non-specific immunity.
b   NO cytotoxicity is chiefly related to cell wall damage.
c   NO is decreased in acute renal failure (ARF).
d   NO is decreased in septic shock.
e   Angiotensin converting enzyme (ACE) inhibitors increase NO release.

15 *Aspirin, NSAIDs, and upper GIT bleeding:*

a   The risk of peptic ulcer bleeding is negligible on very low dose aspirin.
b   Concurrent aspirin/NSAID use multiplies rather than doubles the risk of GIT bleeding.
c   In daily users, the risk of bleeding from aspirin increases the longer the drug is taken.

d   The daily dose of an NSAID bears a direct linear relation to the risk; GIT complications are independent of the patients age.

e   The risk with NSAIDs is greatest when they are taken orally.

16 Selected hypothalamic releasing factors (RFs):

a   Thyrotrophin releasing hormone (TRH) is confined to the CNS.

b   About 75% of brain TRH is found in extrahypothalamic sites.

c   Many patients with depression have a blunted TRH/TSH response.

d   Many patients with depression have increased release of corticotrophin (ACTH) in response to corticotrophin releasing factor (CRF).

e   Some RFs interact with others at the level of the pituitary gland.

17 Some hormonal changes in pregnancy (maternal):

a   LH secretion is increased.

b   Prolactin production increases within a few days of conception.

c   Pituitary growth hormone increases.

d   ACTH and serum cortisol both rise.

e   Adrenal cortex secretes increased renin.

18 Cardiorespiratory fetal and neonatal physiology:

a   In the fetus only 50% of the blood from the superior vena cava enters the right ventricle and pulmonary artery.

b   The left ventricle is thinner than the right ventricle in fetal life.

c   A PaO_2 of 4 kPa, a $PaCO_2$ of 6.5, and a pH of 7.2 are normal for the fetus.

d   Over the first 2 days of postnatal life the $PaCO_2$ rises.

e   The RQ in the neonate is above unity.

19 Pain in the neonate:

a   There are relatively fewer nociceptive nerve endings in the skin of a newborn compared to the skin of an adult.

b   Pain pathways to the brain stem and thalamus are largely unmyelinated at birth.

c   Thalamocortical fibres pain fibres are unmyelinated at birth.

d   Anaesthesia blocks cortisol rise in neonates undergoing circumcision.

e   Giving a pacifier to neonates reduces CVS and RS responses to heel-stick procedures.

20 CNS dysfunction:

a   Glutamate dehydrogenase (GDH) is important for the production of glutamate in the brain.

b   Hereditary spongiform encephalopathies can be infectious.

c   Signs of Parkinson's disease are usually symmetrical.

d The upper body is predominantly involved in arteriosclerotic pseudoparkinsonism.

e Oxidative neurotoxicity from free radicals has been implicated in some neurodegenerative diseases.

Answers

1 a **False.** The Tamm–Horsfall protein is secreted in the kidney by the thick ascending limbs of Henle's loop in response to injury, e.g. damage from Bence Jones protein.

b **True.** Actin is also released in tissue necrosis from any cause.

c **False.** Fibronectins are a family of proteins found in plasma and in the extracellular matrix.

d **True.** One hypothesis for diabetic neuropathy goes as follows: hyperglycaemia $\Rightarrow$ aldose reductase $\Rightarrow$ sorbitol $\Rightarrow$ decrease myoinositol $\Rightarrow$ decrease Na^+,K^+-ATPase.

e **True.** Hence one of the clinical uses for octreotide is in treating bleeding varices.

2 a **False.** Diffusion time increases in proportion to the square of the diffusing distance.

b **False.** The rate of diffusion of lipid-soluble substances across membranes is 1000 to 1 million times slower across membranes than in body fluids.

c **True.** Polar (ionized) substances are lipid insoluble and need special protein channels and/or carriers to cross membranes.

d **True.** Maximal flux cannot be increased once transport proteins are saturated with ligand.

e **False.** It will continue as long as the diffusion gradient exists. The net result is the same as facilitated diffusion but simple diffusion is much slower.

3 a **True.** Medullary hyperaemia (which abolishes hypoxia) washes out the concentrating countercurrent mechanism.

b **False.** The outer medulla may escape damage. Necrosis of renal papillae may develop (furthest from oxygen supply).

c **True.** Other renal vasodilators include NO, platelet activating factor and cytochrome P450 arachidonic metabolites which are all found in the medulla.

d **True.** PGE increases O_2 availability by inhibiting active processes in the thick ascending limb of Henle's loop. NSAIDs inhibit prostanoid synthesis and induce medullary necrosis.

e **True.** Adenosine is released from ATP in hypoxia. Medullary urodilatin (homologous to ANF) may also improve low medullary blood flow.

4 a **True.** To activate prothrombin Xa needs cofactors Va, phospholipid, and Ca^{2+} ions.

  b **False.** Cytokines, glucocorticoids and anaphylatoxins help release + APRs e.g. c-reactive protein, serum amyloid A proteins, clotting factors and inhibit -APRs e.g. albumin, apolipoproten A1.

  c **False.** Macrophages make complement proteins, GIT cells make apoproteins, and many tissues make enzymes.

  d **True.** Albumin is a primary carrier for FFA, bilirubin, and many drugs. It is a secondary carrier for haem, T_4, and cortisol.

  e **False.** Three-fifths of albumin is found in tissue fluid, two-fifths in the plasma.

5 a **False.** The liver weighs about 1.5 kg. Liver blood flow is about 1.5 l/min.

  b **True.** The hepatic artery also maintains the integrity of the connective tissue and bile ducts.

  c **True.** The hepatic artery may provide up to 50% of liver blood flow when the liver is very active metabolically. Normally it only provides 20%, with 80% from the portal vein.

  d **False.** Portal venous blood is not sterile (see hepato-renal syndrome, Chapter 4 Q25).

  e **False.** Portal vein pressure is usually 3–6 mmHg.

6 a **True.** BSP is excreted exclusively by hepatocytes. This test is rarely used nowadays.

  b **True.** Any discrepancy between the volume of blood entering and leaving the liver is due to the large hepatic lymph flow (2 l/day).

  c **True.** Indocyanine green is far less cumbersome to use than BSP.

  d **True.** The liver is vital for maintaining normoglycaemia.

  e **False.** It is almost certain that all plasma albumin is of hepatic origin.

7 a **False.** Several studies have shown that the clinical syndrome of hypoglycaemia bears little or no relation to the rate of fall of blood glucose.

  b **False.** Glucagon provides full defence against hypoglycaemia. Adrenaline is necessary only if glucagon is absent.

  c **False.** The GTT is discredited and is replaced by a 72-h fast test. Whipple's triad consists of hypoglycaemia, hypoglycaemic symptoms, and the reversal of same by glucose.

  d **True.** Hyperglycaemia stimulates aldolase reductase which raises cellular sorbitol. This decreases myo-inositol and causes a decrease in the activity of Na^+,K^+-ATPase (see Q1d above).

  e **True.** Excess glucose is non-enzymatically converted to N-glucosylamine and then undergoes acid–base catalysis to Amadori products. Aminoguanidine blocks degradation of these to AGEs.

8 a **True.** Hypoxaemia and hypercapnia generally occur if $V_d/V_t > 0.5$.

  b **False.** As dead space rises $PeCO_2$ falls below $PaCO_2$.

  c **False.** Over 90% of the cardiac output should participate in gas exchange. The shunt flow (Q_s):total flow (Q_t) ratio should be < 10%.

d **True.** As Q_s/Q_t rises, an increase in fractional concentration of inspired O_2 (FIO_2) produces little increase in the PAO_2 in true shunt although it does improve the PAO_2 in shunt like states (low $\dot{V}/\dot{Q}$).

e **True.** Alveolar gas equation: $PAO_2 = PIO_2 - (PaCO_2/RQ)$ determines the difference in PO_2 between alveoli and arterial blood. Assume that the RQ = 0.8.

9 a **False.** It reduces oedema in this case by preserving ATP.

b **False.** It doubles flexibility of RBCs after about 15 treatments.

c **True.** It helps maintain viability of ischaemic tissues.

d **True.** It is used in necrotising fasciitis, osteomyelitis, clostridial infections.

e **True.** Thus it normalizes local PO_2 and facilitates healing and surgery (including bone grafting).

10 a **False.** In reflexes, they increase the heart rate, the force of contraction, and cause wide vasoconstriction.

b **True.** PGs sensitize vagal afferents e.g. in myocardial infarction. β-blockers inhibit.

c **False.** It is impaired. This may contribute to the excess aminergic stimulation seen in these patients.

d **True.** In this way digitalis may reduce afterload (vasodilation and sympathetic inhibition) (see the Bezold–Jarisch reflex chapter 5 Q336).

e **True.** The release of AVP during fainting or severe haemorrhage may cause bradycardia and vasodilation by this mechanism.

11 a **False.** Depolarization signals the m-gates to open and the h-gates to close.

b **False.** In the closed resting state, the h-gates are open and the m-gates are closed.

c **True.** Depolarization prevents runaway signalling by closing the h-gates. This process is slower in onset than opening of m-gates.

d **True.** Both SA and AV nodes are depolarized by slow opening of Ca^{2+} channels (L-type channels). This accounts for nodal delay.

e **True.** Opening of fast sodium channels also depolarizes general atrial and ventricular mass.

12 a **True.** It usually starts much younger, although F.D. Roosevelt was 54 when his started.

b **False.** Damage is probably due to combination of BP height, neurohumoral factors, and cytokines.

c **True.** This is because of the high prevalence of essential hypertension. However, it is more likely in secondary hypertension from renal disease.

d **True.** Because of this < 1% of primary hypertensives now progress to a malignant phase.

e **False.** Malignant hypertension is more common in lower socio-economic groups, those of African/Caribbean origin, smokers.

13 a **True.** This is no longer seriously disputed. As a rough guide: keep plasma HDL > 1.0 mmol/l, LDL < 3.4 mmol/l, and total cholesterol < 6.0 mmol/l.

b **False.** This has not been proved. Note there is no doubting the benefit of reducing hypercholesteroleamia in patients who have had a myocardial infarction.

c **False.** Several large trials in the mid-nineties show no increase in non-CVS deaths.

d **False.** The efficiency of cholesterol absorption decreases as dietary cholesterol increases. Endogenous production is also inhibited and clearance in bile is increased.

e **True.** Lowering the dietary cholesterol does not work consistently and must be supplemented by cholesterol-lowering drugs if plasma cholesterol exceeds about 8.0 mmol/l.

14 a **True.** Both lung and liver are strategically placed immunological filters. Inducible NO is not Ca^{2+} dependent.

b **False.** NO combines with Fe–sulphur centres of key respiratory and DNA synthetic enzymes in cells. It is also a neuroexcitotoxin.

c **False.** Inhibitors of NO increase and may contribute to hypertension in ARF.

d **False.** NO is increased by cytokines and endotoxin. NO contributes to hypotension, venous pooling, and cardiac dysfunction.

e **True.** ACE (kininase II) inactivates bradykinin which, like other vasodilators, works by releasing NO.

15 a **False.** Even doses as low as 30 mg of aspirin/day are associated with an appreciable risk of haematemesis.

b **False.** The risk of bleeding is roughly added in such a case.

c **False.** Evidence indicates that the risk is much greater in users of less than 1 month compared to longer-term users.

d **True.** Note: the risk also rises with age and varies with the NSAID used (the risk is 10 times greater with azapropazone than with ibuprofen).

e **False.** Suppositories carry a higher risk than tablets. Topical use is also potentially dangerous if plasma levels rise appreciably. Systemic absorption is clearly important.

16 a **False.** TRH has been detected in the GIT, placenta, pancreas, and reproductive tract.

b **True.** However, its concentration is much lower in these sites

c **True.** Many patients with depression also have subclinical hypothyroidism.

d **False.** This response is usually blunted despite hypercortisolism.

e **False.** RFs do not enhance, inhibit, or modify each other's actions at the level of the pituitary.

17 a **False.** Both LH and FSH secretion are inhibited in pregnancy.

b **True.** Oestrogen stimulates lactotrophs; human placental lactogen (HPL) inhibits them.

c **False.** GH is inhibited, probably by HPL.

d **True.** Normal cortisol inhibition of ACTH seems to be inhibited.

e **True.** Renin also comes from uterus, chorion, and kidney.

18 a **False.** At least 97% of blood from the SVC enters the right ventricle.
   b **False.** Ventricles are of about equal thickness.
   c **True.** PaO_2 is 10 kPa in 5 h; $PaCO_2$ is 5.3 kPa by 20 min and pH about 7.39 by 24 h postnatal life.
   d **False.** Over the first 2 days the $PaCO_2$ falls to about 5.5 kPa (respiratory compensation for lactic acidosis of birth).
   e **False.** The RQ is low because brown fat contributes significantly to the infant's metabolism (RQ is 0.7 on day 1 and rises to 0.8 by day 7).

19 a **False.** Density of nocioceptive nerve endings is similar to or greater than that of an adult.
   b **False.** Pain pathways are completely myelinated by 30 weeks' gestation.
   c **False.** Thalamocortical fibres are completely myelinated by 37 weeks' gestation.
   d **False.** Plasma cortisol still rises markedly both during and after the procedure.
   e **False.** But local anaesthesia and/or 50% sucrose applied to the buccal mucosa will markedly reduce such changes.

20 a **False.** GDH breaks down glutamate. Inhibitors of GDH reduce putative neurotoxicity of glutamate.
   b **True.** This occurs via prions (protease-resistant proteins). The first known transmissible spongiform encephalopathy in humans was kuru. Note that this is also the case in Creutzfeldt–Jakob disease and fatal familial insomnia. Note further the strong likelihood that bovine spongiform encephalopathy can transmit to humans and manifest as CJD.
   c **False.** Signs of Parkinson's disease are usually asymmetric. Note that drug-induced cases can be asymmetric too.
   d **False.** The upper body is rarely involved in arteriosclerotic cases, is always involved in primary Parkinson's disease.
   e **True.** An example is dysfunction of superoxide dismutase (free-radical scavenger) in motor neuron disease (role of oxidants in CVS disease also).

Index